[加拿大]沃里克·F. 文森特 著　邓天旸 译

牛津通识读本·

湖泊
Lakes
A Very Short Introduction

译林出版社

图书在版编目(CIP)数据

湖泊 /(加)沃里克·F. 文森特(Warwick F. Vincent)著；邓天旸译. —南京：译林出版社，2021.10
(牛津通识读本)
书名原文: Lakes: A Very Short Introduction
ISBN 978-7-5447-8790-1

I.①湖… II.①沃… ②邓… III.①湖泊-世界-普及读物 IV.①P941.78-49

中国版本图书馆 CIP 数据核字(2021)第 153904 号

Lakes: A Very Short Introduction by Warwick F. Vincent
Copyright © Warwick F. Vincent 2018
Lakes was originally published in English in 2018. This licensed edition is published by arrangement with Oxford University Press. Yilin Press, Ltd is solely responsible for this bilingual edition from the original work and Oxford University Press shall have no liability for any errors, omissions or inaccuracies or ambiguities in such bilingual edition or for any losses caused by reliance thereon.
Chinese and English edition copyright © 2021 by Yilin Press, Ltd
All rights reserved.

著作权合同登记号　图字：10-2018-429号

湖泊　[加拿大]沃里克·F. 文森特／著　邓天旸／译

责任编辑　许　丹
装帧设计　景秋萍
校　　对　王　敏
责任印制　董　虎

原文出版　Oxford University Press, 2018
出版发行　译林出版社
地　　址　南京市湖南路 1 号 A 楼
邮　　箱　yilin@yilin.com
网　　址　www.yilin.com
市场热线　025-86633278
排　　版　南京展望文化发展有限公司
印　　刷　江苏扬中印刷有限公司
开　　本　635 毫米×889 毫米 1/16
印　　张　19
插　　页　4
版　　次　2021 年 10 月第 1 版
印　　次　2021 年 10 月第 1 次印刷
书　　号　ISBN 978-7-5447-8790-1
定　　价　39.00 元

版权所有　侵权必究
译林版图书若有印装错误可向出版社调换。质量热线：025-83658316

序 言

沈 吉

译林出版社出版了一套"牛津通识读本"(Very Short Introductions),邀请我为其中的《湖泊》一书作序。该书作者是加拿大魁北克市拉瓦尔大学生物学教授沃里克·文森特(Warwick Vincent),他担任加拿大水生生态系统研究院首席研究员,也是加拿大皇家学会会员、加拿大皇家地理学会会员,在湖泊研究方面享有很高的国际声誉。他曾于2018年来南京参加第34届国际湖沼学大会,我对他的研究工作印象深刻也非常欣赏,因而欣然答应,也深感荣幸。

1922年,拥有401名建会会员的国际理论与应用湖沼学学会(SIL,又称国际湖沼学学会)在德国基尔成立,这对于湖沼学的发展具有里程碑式的意义,会员几乎覆盖五大洲。此后,每两到三年举行一次学术交流大会。2018年,SIL第34届国际湖沼学大会在中国南京召开,是学会创建96年以来首次在中国举办,由中国科学院南京地理与湖泊研究所承办。我当时作为研究所所长承担了大会的组织工作,并与埃里克·耶珀森(Erik

Jeppesen）教授共同担任大会主席，同时作有关湖泊生态系统长期演化的大会特邀报告。大会期间结识了本书作者。

　　拿到本书书稿，如同旧友新逢，欣然花了两天时间，奉读完成。一个非常鲜明的印象是：作者敬畏先贤，非常推崇弗朗索瓦·A. 福雷尔先生的研究工作。全书通篇穿插福雷尔关于日内瓦湖研究的片段，由此提供湖泊研究的思路、对湖泊认识逐步提高的过程，以及所获得的阶段性成果。洛桑大学（瑞士）的生理学教授弗朗索瓦·A. 福雷尔于1892—1902年间陆续发表了湖泊综合研究专著《日内瓦湖：湖沼学专论》（*Le Léman: Monographie limnologique*）（共三卷），从而宣告了这门学科的诞生。因此，福雷尔也被誉为"湖沼学之父"。

　　本书以讲故事的方式，生动叙述了关于湖泊的那些枯燥的基本概念。比如：通过观察水之运动来阐述波浪和湖流的基本概念；从湖水的季节变化与混合过程的角度叙述温跃层的概念。作者善于对大家日常惯见的现象提出问题，阐述蕴藏于现象背后的科学原理。比如，纯净的湖水为什么是深蓝色？因为水分子能够吸收绿光和红光，后者的吸收程度更高，剩余的蓝光光子则被散射至各个方向并回到我们的眼中，湖水从而显示为蓝色。

　　作者讲述了湖泊中非常有趣的捕食现象，展示了自然界的盎然生机。比如：黄库蚊主要生存于湖泊和池塘，日间潜入水底，以沉积物中的动物为食；夜间它们会迁徙到湖面捕食浮游动物，由此躲避鱼类的捕食。借此，作者介绍了湖泊中非常重要的食物链层次：最下层的被称为生产者，是能够通过光合作用将环境中的无机物制造成营养物质的自养生物，如水生植物和藻类；第二层被称为消费者，指那些以其他生物或有机物为食的异养生物,食草动物为一级消费者,肉食性鱼类为二级消费者;第三层被称为分解者,主要是各种异养细菌和真菌,它们把复杂的动植物残体分解为简单的化合物,最后分解成无机物归还到环境中去,被生产者再利用。

作者非常重视新技术在湖泊研究中的应用,比如同位素方法在食物链研究方面的应用。作者提到,贝加尔湖中浮游植物硅藻在摄取无机氮后,它们的 $\delta^{15}N$ 比大气增加了约 4 ppt[①]。硅藻随后被端足类动物吃掉,并将氮在食物链中一路向上传递,经中上层鱼类(杜父鱼),最终到海豹时 $\delta^{15}N$ 增加了 14 ppt。这种方法对研究食物网中"谁吃谁"的关系提供了宝贵的参考。

作者对于湖泊外来物种的入侵有着十分警觉的意识,列举了当前已经发生的一系列生态灾难。例如,我们较为熟知的原产南美洲的凤眼莲,俗名水葫芦,在亚洲、非洲和美国南部肆虐,将水面覆盖致使水生栖息地窒息,水生生物大量死亡。1968—1975年间,一种叫 Mysis diluviana 的糠虾被引入美国蒙大拿州弗拉特黑德湖上游的三个小湖,以改善鲑鱼渔业。到了1981年,这种糠虾顺流而下进入了弗拉特黑德湖,并于1980年代末在数量上经历了爆炸性的增长。几年后,湖中浮游动物中的枝角类和桡足类因众多糠虾的过量捕食而消失殆尽,淡水红鲑变得没有浮游动物可吃,又因为糠虾只在夜间才出没于浮游区的水面,红鲑无法看见它们因而不能摄食。在弗拉特黑德湖流域,鲑鱼的竞技性捕捞量从1985年的超过10万条直线下滑到1988年的0条。秃头鹰会在淡水红鲑产卵的溪流聚集捕食红鲑,其数量也

---

① 表示数值时,ppt 的规范用法应为 10^{-12},但因书中此类数值出现较多,为了行文简洁美观,沿用原书用法。后文中的 ppb(10^{-9})、ppm(10^{-6})同样如此。——编注

经历了从1980年代早期的600余只到十年后的基本绝迹。

  作者还介绍了湖泊一些引人入胜的奇异现象。比如南极洲麦克默多干谷地区的万达湖，这里的湖面终年结冰，但当第一批科学家在冰面上钻洞，并将热敏电阻探头伸入下方水柱时，他们惊奇地发现温度在随着深度上升，并在底部达到惊人的26℃。又如南极洲西福尔丘的迪普湖，盐度非常高（270 ppt），以至于湖水在冬季也不结冰，可以在周围天寒地冻的湖心划船，而这种液态盐卤湖水的温度达到-18℃。此外，一些长年深埋在冰川下的极地湖泊也令人着迷。

  在最后一章"湖泊与我们"中，作者提醒人们，未来应关注全球变暖对湖泊生态系统的影响。他着重提到，世界各地筑坝拦水，建造了数以万计的人工湖泊-水库，这些水库的负面生态效应已经开始显现；其次是全球范围内湖泊面临的富营养化趋势，这些都是湖泊研究领域亟待解决的世界性难题。

  作为一名湖泊科学工作者，我认为本书是绝佳的科普和通识教育读物。作者在短短的八万字篇幅内，将百年来的湖泊研究娓娓道来，内容丰富有趣，令人目不暇接、心驰神往。湖泊拥有供水、防洪、旅游、灌溉、水产养殖以及维系区域生态平衡等多种功能，被誉为"大地的明珠"。本书引导人们从多学科的视角认识湖泊，并连接湖泊基础研究、湖泊环境治理与管理全过程，必将对推动大众科普教育做出贡献。

中文版序

2018年8月,我有幸受邀前往南京参加第34届国际湖沼学大会。国际湖沼学学会(SIL)致力于内陆水体的研究,自1922年成立以来,每两到三年会在世界不同的地方举办大会。这是大会第二次在亚洲举办,也是第一次在中国举办。

南京的会场坐落于美丽的建邺区,这里有宽阔的马路、闪亮的酒店和有着玻璃外墙的办公楼,有些还在建,在我看来似乎每天晚上都在增高。中国主办方的接待非常出色,我也很高兴能够认识许多来自中国和世界各地的学生与研究者。

值此良机,我们也向我的博士导师,来自加州大学戴维斯分校的杰出的湖沼学教授,时年88岁的查尔斯·R. 戈德曼博士表示祝贺。为向戈德曼教授致意,我们举行了一场特别的研讨会,主题为"世界湖泊给全球的启示",由我的同事,日本湖沼学学会前会长熊谷道夫教授主持。会后我们举办了一场令人难忘的中式晚宴,戈德曼教授在他的友人、同事和曾经的学生的陪同下出席。晚宴由太湖湖泊生态系统研究站前站长、杰出的中国湖沼学家濮培民教授主持。

那个周末，我在汉斯·帕埃尔教授（也曾是戈德曼教授的学生）的带领下对太湖进行了实地考察。我们了解到这一广阔水体作为逾四千万人口水源供给的重要性，以及正在进行的理解和管理水质问题的各项研究，后者面临着全球气候变化带来的额外挑战。

南京之旅使我萌发了将我的湖沼学通识读本译成中文的想法。这本书在2018年经牛津大学出版社出版，后被翻译成法语在魁北克市和巴黎发行。当我听说中译本的出版社——译林出版社就坐落在南京时，我特别高兴，因为这座城市在我于中国介绍湖沼学的过程中有着举足轻重的地位。许丹女士和她在译林出版社的同事在翻译项目上给予了许多鼓励和无价的帮助，我为此向他们致谢。我要特别感谢邓天旸先生，他在拉瓦尔大学攻读化学博士期间勇敢地承担起英译中的工作。作为他博士论文的联合指导老师，我总是担心翻译会占用他研究工作（即开发微流控和高级图谱技术在环境分析系统中的联用）的时间，但幸运的是他能平衡这几项工作，而且我对他在翻译过程中对湖沼学有了深入认识感到非常高兴。

我要向许多湖沼学家致谢，他们在英文原版书出版前后给予了专业反馈和修改意见。他们包括：B. 拜森拿、S. 博尼利亚、R. 科里、A. 卡利、G. 克林、熊谷道夫、U. 莱明、I. 洛里昂、C. 洛夫乔伊、S. 麦金泰尔、S. 马卡格、F. 皮克、R. 皮耶尼蒂兹、M. 劳蒂奥、G. 施拉多、P. 旺罗莱根和A. 维涅龙。我还要感谢阿曼达·托佩罗夫绘制了超高质量的插图，以及加拿大机构对我湖泊研究的资金支持，包括自然科学与工程研究理事会（NSERC）、魁北克自然与技术基金会（FRQNT）、加拿大研究讲座（CRC）、加拿大创新基金会（CFI）、卓越中心网络计划——北极网络计划（NCE-ArcticNet）以及加拿大第一研究卓越基金——北部哨兵计划（CFREF-Sentinel North）。

最后，我想再次感谢南京国际湖沼学大会的主办方，感谢他们的温馨接待以及他们对中国多样的湖泊和河流生态系统的精彩介绍。我希望此书能够帮助环境科学的学生们捕捉他们对湖泊的想象，以及鼓励所有读者去进一步了解世界湖泊水面下潜藏的秘密。

沃里克·F. 文森特
于加拿大魁北克市拉瓦尔大学

纪念丹尼斯·A.沃尔特(1938—2013)

目 录

致 谢 1

第一章 引 言 1

第二章 深邃的水体 6

第三章 阳光与运动 25

第四章 生命支持系统 45

第五章 终端是鱼的食物链 66

第六章 极端湖泊 86

第七章 湖泊与我们 103

索 引 122

英文原文 133

致 谢

作者希望感谢：弗朗索瓦·D. C. 福雷尔同意使用其曾祖父弗朗索瓦·A. 福雷尔的著述中的材料；阿曼达·托佩罗夫绘制了高质量插图；比阿特丽克斯·拜森拿、西尔维娅·博尼利亚、罗斯·科里、亚历山大·卡利、乔治·克林、熊谷道夫、乌尔里奇·莱明、伊莎贝尔·洛里昂、康妮·洛夫乔伊、莎莉·麦金泰尔、斯蒂格·马卡格、弗朗西丝·皮克、莱因哈特·皮耶尼蒂兹、米拉·劳蒂奥、乔弗里·施拉多、皮特·旺罗莱根和阿德里安·维涅龙对书稿提出了宝贵意见；牛津大学出版社的拉莎·梅衣和珍妮·纽吉出色的编辑支持；以及资助作者研究湖泊的机构，尤其是加拿大自然科学与工程研究理事会，和魁北克自然与技术研究基金。

第一章

引　言

什么是湖泊？乍一看，这似乎是个简单的问题：湖泊即被陆地所包围的水体。但这个枯燥死板的物理定义仅仅是回答的开端，湖泊的本质和含义有着许多其他有趣的解读。对淡水生物学家来说，湖泊就是陆地上的绿洲，其中微生物、植物和动物相互影响，湖泊内的物种、食物网和生态过程也亟待探索。而环境学家则从化学角度出发，将湖泊视为和大气层交换气体的活反应堆。湖泊收集并转化从邻近集水区冲洗出来的物质，同时也是水生植物和藻类通过光合作用合成新的有机物的场所。我的一些同事从事湖泊沉积物中微观化石的研究，对他们而言，湖泊是丰富的信息库，可以告知过去和现在，并引导我们制订未来的计划。

对水文工程师和大众来说，湖泊是不可或缺的资源，通过管理、治理现有湖泊和人工造湖，可以满足日益增长的饮用水、水力发电、渔业和其他生态系统服务等需求。维持这些服务需求则需要对表面水和地下水的平衡保持密切关注，包括它们的流入量、蒸发量、抽取量和流出量,这些条件共同决定了湖盆中保有的水量。水资源在世界上的许多地方都极度匮乏。在当前全球气候变化的大环境下,湖泊的水量出入平衡越发岌岌可危,其管理也显得困难重重且富有挑战。

从物理层面来讲,湖泊系指由太阳与风所驱动的持续运动的水体。在不同的季节,湖水可依据许多不同性质的差异分层,如温度、含氧量、颜色和含盐量。这些分层有时会出乎意料地显著,但在每年的混合期则会被打破。湖泊是陆地景观中水体联通且缓慢流动的管道:水从其中流入流出,但这种有序运动会持续地被风力引起的漩涡、流涡和逆流所干扰,甚至在水下,这种波的生成和破碎也在进行。

每年夏天,当我前往科考场所,飞过加拿大北部地区上空的时候,目光所及之处,下方的淡水湖犹如闪闪发光的群岛;或在更北更寒冷的纬度,白雪覆盖的湖冰镶嵌在起伏的苔原上。在部分研究中,我们关心微观生物在这些如群星般繁多的极地湖泊和池塘中的分布,以及这些具体的湖泊或池塘是怎样决定了这种分布。而对于研究鱼类和其他水生生物进化的人来说,最古老的湖泊群即一个个实验室。其中发生的繁殖、基因变异和物种形成的过程,能够帮助我们理解这个星球上的生物多样性是怎样进化而成和持续变化的。查尔斯·达尔文甚至推测生命可能是在"一些温暖且富含氨和磷盐的池塘"中诞生的。

湖泊是陆地的最低点,并最终汇入海洋(除有些例外)。从这个角度看,湖泊作为其周围环境的集成者(图1),反映了其流域(又称集水区或水域)供水、植被、地质、环境内的自然活动和人文活动的综合影响。所以湖泊也可作为反映环境变化的指

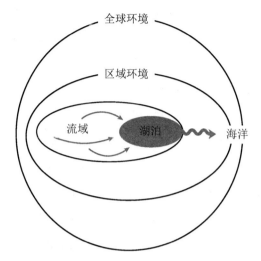

图1　湖泊是环境的监测者、集成者和连通者

标,用来监测局部环境中人文活动的当前规模与长期影响,以及区域和全球范围内正在发生的气候变化、污染物扩散和生物多样性的变化。

接下来就是面积的问题。达尔文提出的那个温暖的小池塘能被视为湖泊吗?有的人将池塘定义为深度允许人蹚过的水体,但在沼泽湿地,这种以身试深的测试方法并不明智。有的人将池塘定义为可被冻结至水底的水体,而湖泊则不能,但这种十分具有加拿大特色的对水生世界的认知却并不适于其他地方。此外,即使在加拿大,表面覆冰的底层水也特别抗冻。在英格兰湖区,游客会发现当地人把小型水体称为"山中小湖"(tarns),大一些的则被称为"湖"(lakes)、"池子"(meres)或者"水"(waters),对具体称谓的使用并没有明确的共识。文学中北美洲最有名的湖泊是瓦尔登湖(Walden Pond),纽芬兰的人们把大部分的湖都叫作"池"(ponds),包括长达16千米深165米的西布鲁克池,这进一步加深了称谓不明的迷惑性。所以,"湖"和"池"最好一并讨论,而在泛指各类水体时则使用"湖泊"一词。

无论我们想统计世界范围内,还是具体某个国家,抑或我们所处的周边地区的湖泊数量,湖泊的面积大小都非常重要。我们需要设定湖泊面积的最小成像阈值去界定湖泊。而随着卫星乃至无人机遥感技术的进步,这一阈值逐年下降,湖泊存量的下限也是如此。高分辨率卫星能够很容易地分辨出0.002平方千米以上的湖泊,这等同于直径约为50米的圆形水域。依此标准,研究人员从卫星图中统计得出全世界约有1.17亿个湖泊,总面积达500万平方千米。加拿大人认为本国是世界上湖泊最多的国家,包括和美国共有的五大湖群,并且保有地球20%的地表淡水量。但我办公室墙上的世界地图总提醒我:俄罗斯也是一个幅员辽阔且湖泊众多的国家,其中的贝加尔湖是世界上最深的湖泊,同时它也保有地球另外20%的地表淡水量。

此书旨在对湖泊相关科学知识作简短的概述,包括它们作为我们所属并依赖的生态系统的功能,和它们对环境变化的反应。当然湖泊对人类探索和文化的重要性,还有其他超越科学但同样多样的原因。深不见底且幽暗的外观赋予了湖水神秘莫测和令人不安的气质,一直深深吸引着小说家和诗人就此进行创作,如西尔维娅·普拉斯的《涉水》和威廉·华兹华斯的《序曲》,后者描写了他在孩童时于午夜涉过一片幽暗可怖的湖的旅程。湖泊深处栖息着神话生物的传说在世界各地的文化中流传,如毛利人神话中住在江河湖海里的塔尼华水怪。湖泊有着更深远的精神意义。在玻利维亚和秘鲁的神话中,的的喀喀湖湖底的太阳神因蒂，养育了印加文明的创建者——曼科·卡帕克和玛玛·奥克略。湖泊如同宁静的镜面和色彩缤纷的调色板，除了吸引许多旅客每年前往湖岸边度假，也一直是各个文化中艺术家、音乐家、作家的灵感源泉。有的作品，如松尾芭蕉的经典俳句，能够唤起读者听觉上对湖泊和池塘的想象（"水声响"）；在加斯通·巴舍拉关于梦境的著作中，他分析了为何池水、泉水、湖水和溪水是"物质想象"和遐想的基本元素。

本书将把重心放在科学层面。而我希望这本通识读本可以让读者在下一次游览湖泊的时候，能以一种更强烈的好奇心和求知欲去认识湖泊，去认识水面和水下的奇观。每个章节将简短地介绍有关湖泊物理、化学、生物特征的概念，同时着重阐述这些特征之间如何相互作用，湖泊和人类的需求之间的关系，以及全球环境变化对湖泊的影响。湖泊科学有着悠久的观察史和发现史，也有许多优秀的教材在更为学术的层面介绍湖泊生态。这些书本包含众多既有理论和新的信息，但所有这些知识都可以溯源到19世纪，一个年轻科学家在瑞士阿尔卑斯山脚美丽的日内瓦湖畔做出的决定。

第二章
深邃的水体

> 我面临着两种选择：要么建立自己的实验室，研究解剖学、组织学和生理学这些我在自然科学院不得不教授的学科……要么把这个湖当作我的实验室和水族馆，它充满神秘并吸引我去研究它。我很快做出了决定……
>
> F. A. 福雷尔

作为一个新受训的科学家和医学博士，回到日内瓦湖畔后，弗朗索瓦·A. 福雷尔选择了一条职业道路。这条职业道路将引导他接下来数十年的研究，并最终为现代湖泊科学奠定了基础。日内瓦湖位于法国和瑞士的交界，而福雷尔则在瑞士一侧的湖畔出生长大。他敏锐地意识到这一湖泊对当地居民广泛的重要性。首先，日内瓦湖是当地社区首要的饮用水水源，包括正蓬勃发展的洛桑市。福雷尔正是在此处的学院（今洛桑大学）被任命教学。

福雷尔在儿时常常随父亲去日内瓦湖（法语称之为莱蒙湖）探索近岸水面下古老的高脚屋村落。这处地点激发了小福雷尔的想象。这些青铜时代的聚居地遗址埋在水面下,在此处发掘的考古文物充分证明了日内瓦湖和人类源远流长的关系。福雷尔也认识到日内瓦湖在渔业和运输业上巨大的商业价值,并随后用经济学的知识对这些价值作了定量分析。同时他很欣赏日内瓦湖和周围群山展现的美学魅力,并乐于和风景画家弗朗索瓦·L. D. 博西昂结伴同行。但他尤为感兴趣的是日内瓦湖那深邃碧蓝的湖水下潜藏的科学秘密,其中许多奥秘在经过严谨的研究后都能加以揭示。

1867年,在瑞士、法国和德国经过约11年的学习和医科培训后,26岁的福雷尔回到了洛桑,住在他靠近莫尔日的祖屋。他"全方面地研究湖泊"的决定一开始让他之前的德国导师有些担忧,后者让他采取一条更加细分的路径。福雷尔不为所动,并从各个学科方向对湖泊开展研究:从波浪、水流、阳光的穿透和水中化合物的性质,到生活在湖泊各个角落的植物、动物以及微生物。不过直到许多年之后,他才把这些分散的研究整合起来。

1892年,他出版了关于日内瓦湖的综合性著述,在其第一卷的序言中,他借用湖泊的希腊语 *limne* 一词创造了湖沼学(limnology)这个术语,并把这门新的综合性学科定义为"关于湖泊的海洋学"。今天,湖沼学已经把研究对象拓展到了河流、湿地乃至河口,但还是着重于研究湖泊和池塘。湖沼学研究人员在世界范围内有很多学术组织,如国际湖沼学学会(SIL)及湖沼学和海洋学学科联盟(ASLO)。然而湖沼学作为一个科学术语在该领域以外并不为人所知。英语中形容淡水水体的词汇 lake 并不是从一个相似的词根衍生而来(如海洋学 oceanography 源于希腊语 *okeanos*),而是从拉丁语中的盆地 *lacus* 而来。另一方面,湖沼学的概念直观且有吸引力,而且和我们当下对湖泊在保护、恢复和管理方面的目标高度相关。

对福雷尔来说,湖泊科学可以细分至不同的次级学科和课题,所有这些课题至今还在吸引淡水科学家们的关注。首先,湖泊的物理环境包括其地质学起源和背景、水平衡及与大气的热交换、光的穿透、温度随水深的变化、波浪、水流以及共同决定水运动的混合过程。其次,化学环境也很重要,因为湖水含有丰富多样的溶解物质(溶质)和颗粒物。这些物质在湖泊生态系统功能中起着不可或缺的作用。最后,湖泊的生物学特征不仅包含植物、微生物和动物的单个物种,也包含了它们构建成的食物网,以及这些群落在湖底和湖面的分布及功能。

福雷尔的新学科体系中有两点让他的思想从当时众多的学者中脱颖而出,甚至超越之后的许多专家。正如他在湖沼学最

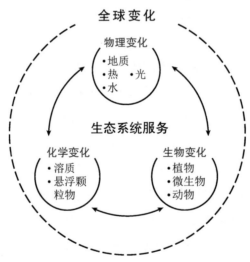

图2 全球环境变化下对湖泊生态系统服务产生影响的交互作用

早的定义中所提出的,从不同角度研究湖泊能够实现知识之间的相互交流并揭露其中的联系,从而绘制出生态系统的全景图。湖泊的物理、化学和生物特征之间的相互作用尤为引人注目。例如,他说明了陆地的有机物是如何引起湖水变绿并影响水体透明度的;又如周围流域中岩石的风化如何影响了湖泊中的矿物质;再如栖息在湖底的动物与上层湖水中浮游生物生长和死亡的关系。和当时把湖泊视为封闭微观世界的流行观点截然不同,他在1891年时写道:

> 实际上,通过和表层大气持续的气体交换,通过流出水体带走溶质和不溶物,通过支流引入新的物质,湖泊在和世界其他地区交流。

第二点涉及人类。福雷尔认识到湖边的居民实际上是日内瓦湖生态系统的一部分,他们在许多方面都仰仗于湖泊提供的服务,如安全的饮用水、商业性渔业、承担客运和货运的运输水道、在湖边生活的审美乐趣和心理健康。福雷尔观察到,这些价值因人类活动而日趋减少,如水位管理不力、入侵性水生物种的引入以及由下水道进入湖泊的人类病原体污染。人类作为生态系统一部分的观念在大半个20世纪内都未得到充分认可。随着全球变化的影响逐渐加深,以及维持我们从属并赖以生存的生物圈充满挑战,这一观念在今天尤为重要。

湖泊的诞生与死亡

福雷尔在他关于日内瓦湖的三卷本著作的第一卷中花了

大量篇幅讨论湖盆可能的起源。他也描述了沉积物在湖中的累积方式，尤其是那些来自源于冰川的罗讷河上游的沉积物，它的河水浑浊且富含矿物质颗粒。沉积物或由入流引入，或由湖中的微生物析出，在湖床上持续累积。这意味着湖泊是地表景观中短暂的存在。自其诞生起，湖泊就不断被填充直至逐渐消失。世界上最深最古老的淡水生态系统，位于俄罗斯西伯利亚的贝加尔湖就是一个引人注目的例子：其湖水最深可达1 642米，但水底下方的湖盆更深，在它超过2 500万年的地质历史里填充了7 000米深的沉积物。

　　湖泊的起源多种多样：由地壳运动构成的构造湖，冰川侵蚀或退化形成的冰川湖，河流冲刷而成的河成湖，火山口湖，河边湖，陨坑湖和其他类型的湖泊，包括人类建造的池塘和水库。构造湖可能由单个断层构成，如贝加尔湖和坦噶尼喀湖（东非），也可能由一系列交错的断层构成。例如，太浩湖（美国）因断层阻塞，有着长方形马槽状的湖盆，平均湖深300米，最深可达501米。世界上最古老最深的湖泊大多都是构造湖。其悠久的历史让地方特有的植物和动物能够在此演化，换句话说，这些物种只能在这些地点找到。

　　构造湖和其中地方特有的动物群的例子以东非大裂谷中的湖群最广为人知。这些湖的湖盆已经隔离了足够长的时间，能让其中无数鱼类呈现适应性辐射的现象。马拉维湖有着种类最多的鱼，共超过850种，其中大部分都是地方特有的，分布在11个科下，以丽鱼科最多。在坦噶尼喀湖，共有16个科的鱼类，丽鱼科就有200种。而在维多利亚湖的广阔水域中（6.88万平方千米，最深可达84米），曾一度栖息着超过500种鱼。这些湖都面临着来自农业发展、渔业捕捞和物种入侵的压力。例如，尼罗河尖鲈被引入维多利亚湖后，由于捕食和竞争的增加，以及水质的改变，导致可能多达200种地方特有的物种灭绝。

其他具有地方性鱼类的古老构造湖还包括日本的琵琶湖，有17种地方特有的鱼类，如琵琶湖巨鲶（*Silurus biwaensis*）；的的喀喀湖栖息着15种地方性鳉鱼（*Orestias*）和一种巨型水生青蛙（*Telmatobius culeus*）；奥赫里德湖有着地方特有的海绵动物和50种腹足类动物；贝加尔湖容纳着超过1 000种地方性物种，包括浮游植物如硅藻属的贝加尔浮生直链藻，无脊椎动物如端足类、浮游动物优势种贝加尔侧突水蚤，鱼类如贝加尔杜父鱼以及唯一一种淡水海豹贝加尔海豹。

就总数而言，世界上绝大多数的湖（包括图3中英格兰湖区的湖泊）都起源于上一个冰河时期：冰川从岩石中凿开湖盆，加深了河谷。这些湖泊包括：欧洲大陆的深水湖如日内瓦湖（310米深）、博登湖（251米深）、马焦雷湖（372米深）和科莫湖（425米深）；苏格兰的淡水湖如尼斯湖（227米深）和莫勒湖（310米深）；北美洲的大湖如密歇根湖（281米深）和苏必利尔湖（406米深）；新西兰南岛的湖如瓦卡蒂普湖（380米深）和豪罗科湖（462米深）。这些冰川运动也在地面上造成无数较浅的凹陷，如在加拿大北部的前寒武纪花岗岩上遍布着只有几千年历史的湖泊，许多这些年轻的极地水体底部只有浅浅的一层沉积物。

随着冰川的后退，其末端冰碛（堆积的砾石和沉积物）在地表筑坝，提升水位或创造了新的湖泊。这些冰碛拦截而成的冰川湖包括智利南部湖区那些美丽的湖泊，如延基韦湖（317米深，也受火山活动影响）和里尼韦湖（323米深）。南美洲最大

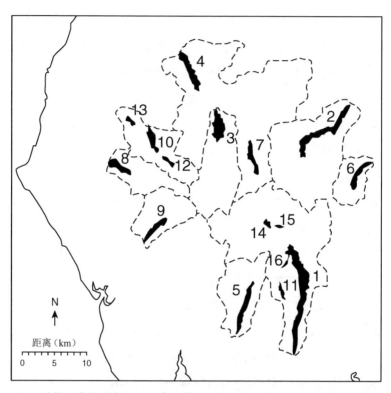

图3 英格兰湖区的湖泊以及其流域。英格兰湖区（表格1）坐落在英格兰西北部，自1920年代以来就是英国淡水生物协会（FBA）进行淡水研究的主要场所，近来则由水文和生态中心（CEH）主导研究。这些湖泊呈轮辐状分布，这种分布形状被认为来自中央穹丘上的放射状排水系统，在冰川时期中央穹丘和山谷被冰川腐蚀和加深

（1 850平方千米）最深（586米深）的冰碛截坝而成的湖有两个名字，因其在巴塔哥尼亚地区而横跨两个国家：在阿根廷被称作布宜诺斯艾利斯湖，而在智利被称作卡雷拉将军湖。在世界上许多地方，随着冰川后退，大块的冰从冰川中断裂开来，被留在冰碛中。它们随后融化并填充了湖盆，被称为"壶穴湖"或"锅状湖"。这种湖在北美洲和欧亚大陆的平原广为人知。

表格1　英格兰湖区的湖泊（编号与图3对应）

编号	湖泊	面积（km²）	最大湖深（m）	平均湖深（m）
1	温德米尔湖	14.8	64.0	21.3
2	阿尔斯沃特湖	8.9	62.5	25.3
3	德文特湖	5.3	22.0	5.5
4	巴森斯韦特湖	5.3	19.0	5.3
5	科尼斯顿湖	4.9	56.0	24.1
6	霍斯沃特水库	3.9	57.0	23.4
7	瑟尔米尔水库	3.3	46.0	16.1
8	恩纳代尔湖	3.0	42.0	17.8
9	沃斯特湖	2.9	76.0	39.7
10	克拉莫克湖	2.5	44.0	26.7
11	埃斯韦特湖	1.0	15.5	6.4
12	巴特米尔湖	0.9	28.6	16.6
13	洛斯湖	0.6	16.0	8.4
14	格拉斯米尔湖	0.6	21.6	7.7
15	莱达尔湖	0.3	20.0	7.0
16	布莱勒姆冰斗湖	0.1	14.5	6.8

随着冰川和冰盖在地表上摧枯拉朽式地推进，它们的终点可能就是被冰川冰流不断向前推进的融雪湖。当冰川退化，这些"冰前湖"则可能急剧扩张，直至它们的冰坝缩小或被冲破。其中最壮观的例子当属北美洲的两个冰前湖：阿加西湖和欧及布威湖，于上一个冰川期形成于劳伦特冰川前。前者的面积在1.3万年前达到最大值，约为44万平方千米，几乎是现今北美洲五大湖群的两倍。大约8 200年前，在位于北哈德逊湾的一块冰盖坍塌后，湖水涌流而出，已连成一片的阿加西–欧及布威湖几

乎完全干涸，并使全球海平面上升了超过0.8米。这一灾难性的事件引起了当时海洋洋流和气候的急剧变化，并随之改变了人类迁徙的规律和欧洲的史前农业文明。

最激烈的湖泊形成过程都是火山运动的结果。火山喷发后留下的坑洞被水填充，形成了小规模酸性湖。这种湖泊往往是圆形的。世界上海拔最高的湖泊位于智利和阿根廷边界的奥霍斯-德尔萨拉多活火山口，是一个海拔高达6 390米的小型水体。更大面积的火山湖是在爆发后坍塌的岩浆房中形成的，被称为破火山口湖。已知最大的火山口湖是陶波湖（现有616平方千米，深186米），由26 500年前新西兰北岛中部的超级火山爆发形成。这次爆发喷射出了超过1 000立方千米的物质，其引发的山体坍塌形成了一个巨大的破火山口，并在水填充后形成了湖泊。最近的一次后续喷发发生在约5 000年前。而今天，湖泊内部和附近地区的地热活动则证明，这一地区的地质活动一直处于活跃状态。

小行星撞击而成的陨坑也能够成为湖泊的湖盆。这些湖泊有着极大的科学价值，也吸引了许多人的关注。最有特色的一个例子是曼尼古根湖。这是一个大型（占地1 942平方千米，深350米）环状湖泊，坐落于加拿大魁北克市的中部，由2.14亿年前的一颗直径5 000米的小行星撞击而成。平圭勒湖位于魁北克更北部的亚北极地区，有着近乎完美的圆形轮廓，直径为3 000米（图4）。因纽特人很早就知道了这个湖泊，他们相信水晶般清澈的湖水充满了治愈的力量，于是称之为"努纳维克的水晶眼"。其所占据的陨坑于1 400万年前由小行星撞击而成。由于陨坑极深（400米深，现湖泊深267米），平圭勒湖的水体在冰

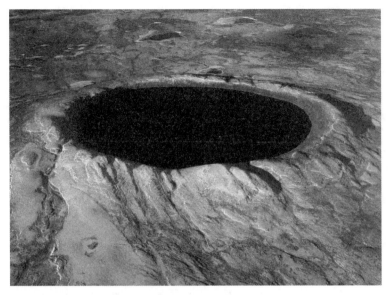

图4 平圭勒湖,北魁北克的"水晶眼"

河时期的冰川厚冰下很可能没有冻结,而成为冰下湖,或许和今天在南极发现的那些冰下湖(沃斯托克湖、威兰斯湖、埃尔斯沃思湖)情况类似。湖沼学家从湖底深厚的沉积物中采样,以提供在几个冰期—间冰期循环过程中不受干扰的气候记录。

曾几何时,湖沼学家对小型湖泊和池塘不甚关注。然而当人们意识到这些规模并不起眼的小型水体在数量上极度丰富,并占据了一大部分地表面积时,这种忽视的态度就转变了。此外,这些小型水体往往有着较高的化学活性,如温室气体排放和营养物循环。同时,它们也是包括水禽在内的多种动植物的主要栖息地。一个典型的例子是北极的解冻湖("热喀斯特湖"),由富含冰的永冻土壤(永久冻土)解冻形成。这些湖泊大量出现在北方地表景观中,总面积超过25万平方千米。由于全球气候变化,永久冻土的融化和退化让这些小型水体经历着激烈的变化。在一些地方,它们随着蒸发、填充或排水而消失,而在其他地方则在大小和数量上都有增加。这些水体也是微生物活动的热点,微生物将之前储存在冻土里古老的碳转换为二氧化碳或甲烷,然后释放到大气中。

湖泊的水下形状

表现湖盆形状的三维形式或形态测量的等深线图是研究任何湖泊必不可少的第一步。尽管现在大部分的等深线图都已经数字化了,但世界上仍有部分湖泊没有这一基础数据。一旦有了等深线图,几个重要的数值就可随之计算出来。首先计算的是每个等深区间的面积,往往通过地理信息系统(GIS)的软件集成包就能完成。这些面积随后可以作为深度标绘在图表上。这一面积-深度图被称为陆高水深曲线,便于确定一些有用的统计数据。

看到贝加尔湖的曲线图(图5),我们会问:这个古老的湖有多大部分湖深超过500米?对于世界上绝大多数的湖来说,答案是没有,因为它们的最大湖深也比这个小得多。但是贝加尔湖的等深线图显示它有三个深湖盆,陆高水深曲线则将复杂的形态测量结果转换为一条简单的曲线,揭示多达68%的湖面下水深都超过500米。同样,我们也可根据这一曲线很快看出,有50%的湖面下水深至少达790米。陆高水深曲线上不同深度的数值可以相加以计算湖的容积,对于贝加尔湖来说,这一数值为2.3万立方千米,可将英格兰淹至水面176米以下。湖的容积除以湖的面积则能给出湖沼学上的另一参数,即平均湖深。贝加

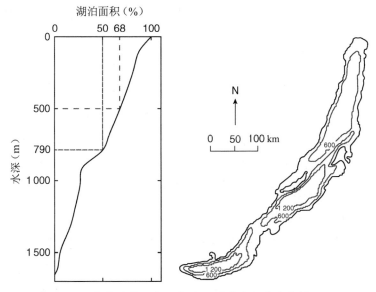

图5 俄罗斯贝加尔湖的等深线图（右）与陆高水深曲线（左）

尔湖的平均湖深有744米。通常来说，平均湖深越高，湖水的透明度就越高，水质也越好。但人类活动会对这一特征造成严重破坏，一如在贝加尔湖所观察到的那样。

水位的起伏

用最简单的水文学术语来说，湖泊可以被看作地表景观上的水箱，不断地由入流河补充水分，而通过出流河将多余的水引走。基于这一模型，我们能提出一个有趣的问题：水分子在流出前平均会在这个湖里停留多久？这一时间值被称为水滞留时间，用湖的容积除以出水口的径流量即得。这一参数也被称为"冲换时间"（或"冲换速率"，假如以单位时间内排出湖容积的百分比计算），因为可以用于估算矿物盐或污染物流出湖盆的时间。一般来说,湖泊的冲换时间越短,其对流域内人类活动造成影响的抵抗力就更强,虽然不会完全免疫。

每个湖泊都有自己独特的流域规模、容积和所处气候,这些因素综合起来让不同湖泊的水滞留时间有着巨大的差异(图6)。例如,作为魁北克市饮用水源的圣查尔斯湖由河床截坝而成,由一个相对湖泊面积(3.6平方千米)较大的流域(169平方千米)灌溉,因此视季节有一至数月的水滞留时间。另一极端是的的喀喀湖,只有相对于容积很小的排放流量,因此其估算的水滞留时间超过1 000年。

另一种更为精确的滞留时间计算方法是考虑这个问题:假如把湖水抽干,需要多长时间才能重新充满?对于大部分的湖泊来说,这一方法所得结果和通过出流量计算的方法所得接近;但对于通过蒸发保持水平衡的湖泊来说,所得结果则要短得多。而的的喀喀湖就属于这种情况,其基于入流量计算的实际滞留时间只有80年(而不是基于出流量所得的1 200年,见图6),因为通过蒸发损失的水比径流损失的更多。缺乏完全冲换过程也

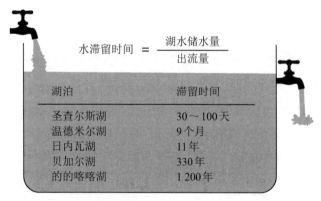

图6 不同湖泊水滞留时间的差异

意味着盐的浓度更高，水也微咸（盐度约为0.7‰）。

假如进入图6水箱中的水和从出水龙头离开的水总体积一样，那么水箱中的水位就会保持恒定。但对于湖泊而言，情况往往并非如此。住在湖边的人们在暴雨倾盆或是其他事件过后，会看到湖水水位急促甚至可怕地波动。一个极端例子是智利南部的里尼韦湖：1960年的一次强烈地震引发的山体滑坡堵住了出流口，导致水位上升20米，下游的城市瓦尔迪维亚及其周边地区不得不计划撤离10万人，以应对水位突破堤坝后将导致的严重洪灾。还好接下来的几周内，通过挖掘泄洪通道，人们以可控的方式将堆积的水引出。

河流流量的自然循环也能引起湖泊的巨大变化。亚马孙河及其广阔的洪泛平原（葡萄牙语中被称作 *varzéa*，意为"低洼地"）就是最明显的例子。许多鱼类都依赖于亚马孙河的年度行洪循环，以便溯流至沼泽森林以陆生昆虫、蜘蛛、坚果、种子和花朵为食。卡拉多潟湖是亚马孙雨林腹地中众多湖泊中的一个，位于马瑙斯——一个地处内格罗河和苏里摩希河交界的巴西城市——上游60千米。如同该区域的其他湖泊，卡拉多潟湖形状交叉复杂，当湖盆被卡布奇诺色的苏里摩希河水灌满，湖的水位会上升10米，并且面积扩大至原来的4倍。低洼地湖大多都长有浮草，如双穗雀稗和多穗稗。这些漂浮的草甸随着季节变化生长枯荣，给这些草岛上的昆虫、鸟类和蛇提供了栖息地。

气候变化通过改变入流量及蒸发损失的平衡，对湖泊水位起着主导作用。最瞩目的例子当属撒哈拉沙漠南部边缘的乍得湖。由于水浅（最大深度11米，平均深度1.5米），乍得湖对季节性或长期的降雨变化高度敏感。在过去的50年中，由于气候日

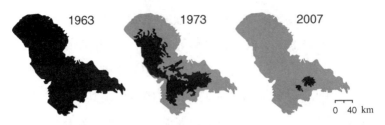

图7 持续干燥环境下中非乍得湖的萎缩

20 趋干燥,同时又受到截坝、灌溉和围湖造田这些无效人类活动的影响,湖泊的面积急剧收缩。渔民和农民的暴力冲突时有发生,因为两者对水的需求截然相反。地质记录表明,乍得湖在过去经历了深刻的变化：从曾经超过100万平方千米的大乍得古湖,到如今湖水几近干涸。现在,超过3 000万人都依赖湖水生活,而他们正面临彻底丧失这一资源带来的灭顶之灾。

湖水水位下降也能带来一些惊喜。以色列的基尼烈湖（面积167平方千米,最大湖深43米）也被称为加利利海,在《圣经·新约》中有重要的地位。在1980年代后期的一次干旱中,这一淡水湖的水位下降了9米,显露出一处石器时代的椭圆形小屋聚居地（这一考古遗址现在也被称为奥哈洛二号遗址）,可追溯至2.3万年前。这是世界上最古老的人类聚居地之一,也是人类和湖泊长期共存的证据。

湖泊沉积物——历史档案馆

21 乍得湖是湖泊受环境变化影响的一个极端例子,但即使气候最细微的变化和人类最轻微的活动也会被记录在湖中,只要经过仔细分析便可获取其中奥秘。每一年矿物和有机物颗粒都会随风沉降在湖面,或从流域被冲刷到湖中,同时湖泊中的水生植物或浮游植物不断地生产有机物。这些物质不断沉降，最终在湖床上累积成一年份的沉积物层。这些沉积物是信息库，记录着周围流域过去的变化，并长久记载着在湖沼学上湖泊对这些变化的反应。分析这些自然档案馆的工作被称为古湖沼学（在海洋研究中则被称为古海洋学），而水科学的这一分支为湖泊如何随时间变化提供了许多见解，包括污染物的出现、影响和消失，湖泊内外植被的变化，以及区域性和全球性气候的改变。

古湖沼学的采样过程都是从抽取湖泊沉积物填充采样柱开始的。有许多类型的便携采样仪器可以抽取少量浅层沉积物以分析近几百年的记录，而更大型的钻探设备则可以采集更深的沉积物，把研究范围扩展至数万年乃至更久。比如西伯利亚的埃利格格特根湖是由一颗小行星在358万年前撞击而成的，从其中采得的一根长达400米的沉积物和岩石柱记录了过去360万年内北极气候的连续变化，包括上新世到更新世的转变。在日本由300万年前的地壳运动构造成的琵琶湖，一根14 000米采样柱中上端的250米的湖泊沉积物，则提供了可以追溯到43万年前的记录。

古湖沼学的研究采样一般是在湖泊最深处进行的，以获得湖盆更全面更完整的历史。这些地方通常也是沉积物最多，受底栖动物干扰（即沉积物的生物扰动）最少的地方。在采样柱被带到水面后，其中的沉积物样品将被岩心桶挤出并分段。上层沉积物的年代一般通过测量放射性同位素铯-137和铅-210含量，可以确定至约150年前；而更深更古老的沉积物则要通过测量放射性同位素碳-14确定。通过这几种放射性同位素测定法，辅助以插值法，每一层沉积物的年代便可估算。这些标记了年代的沉积物层将被进一步分析,作为变化证据。

从那些采样柱切层而得的样品在显微镜下乍一看只不过是一张毫不起眼的涂片,上面尽是些尺径和形状都随机分布的颗粒。然而,通过仔细的观察,我们还是能够在这些颗粒中分辨出那些尚未分解的陆生和水生生物的残骸,而很多可以识别出它们的起源物种。这些微型化石包括被冲入或吹入湖泊的花粉粒,它们因为有坚硬的外壳而未被分解。因为这些花粉独特的外形,它们可以被识别到属甚至具体到物种,因而湖泊的沉积物记录着周围地表环境植物群结构的变化。

湖泊沉积物中信息量最丰富的微型化石是硅藻。这是一种藻类,其细胞壁由硅玻璃组成,可以抵抗分解。每个湖泊基本上都含有数十到数百种的硅藻,每种都有自己独特的环境偏好,每种硅藻独特的细胞壁形状和装饰也便于我们在沉积物的微型化石残迹中将它们识别出来。一种广泛采用的做法是在多个湖泊中采样,分析表层沉积物中硅藻的种群组成,再在它与湖水的某些变量(如温度、pH值、磷含量或溶解有机碳含量)之间建立统计学关系或某种"传递函数"。这种定量的种群-环境对应关系可以应用于同一区域湖泊采样柱中每层的硅藻化石种群组成,这样就可以逐年重建和追算湖泊过去所经历的物理化学变动。其他用于推断环境变化的化石指标包括藻类色素、细菌和藻类(包括引起有毒水华的物种)的DNA以及水生动物的残骸,如介形虫、枝角动物和昆虫的幼虫。

挖掘这种地理历史档案的一个例子是对瓦尔登湖的研究(图8),它是一个坐落在美国马萨诸塞州波士顿附近的壶穴湖,占地25公顷,最大湖深约为30米。这一湖泊在美国文学中享有盛名,因为自然主义者、散文作家、哲学家和历史学家亨利·戴维·梭罗从1845年7月4日到1847年9月6日在这里居住了两年。他在1854年出版了《瓦尔登湖》,将他在这里居住的经历和对自然的思考撰写成经典著作。他写道:"湖泊是自然景色中最美也最富有表现力的特征。它是地球的眼睛;凝视湖中,人能够衡量出自己本性的深度。"

梭罗在他的日记中详细地记录着许多湖泊的特征,包括湖水上暖下冷的分层。这比福雷尔开始研究日内瓦湖早了大约20年,比康奈尔大学的湖沼学教授詹姆斯·G.尼达姆出版第一部淡水生态学英文教材早了半个世纪。尽管梭罗自己并不愿意接受科学的各个方面,这不妨碍他被视作北美洲的第一位湖沼学家。

图8中的花粉信息由一根28厘米长的沉积物芯给出。野草花粉的增加说明橡树的砍伐和农业用地的扩张,这些变化见证了新英格兰早期殖民者的垦荒。讽刺的是,在梭罗体验湖泊和森林的精神价值同时,全面破坏森林的规模达到了顶峰,80%的土地被砍清,转换成了农业用地。到20世纪早期,花粉记录中的这一过程发生了逆转:因为住在乡村的人们因工作搬进城市,

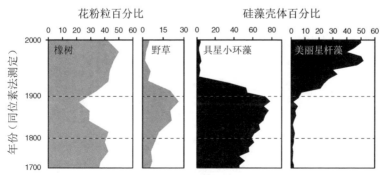

图8 美国瓦尔登湖的沉积物对近300年来环境变化的记录

让没有经济价值的耕地重新变成了森林。沉积物中野草花粉的减少和橡树花粉的对应增加说明了这一点。

从对硅藻的分析可知，瓦尔登湖自1880年至今还经历了其他的一些变化（图8）。某些硅藻，如具星小环藻（*Cyclotella stelligera*），先是逐年增加成为显著种，然后突然被一些富营养水体的特征藻取代，如美丽星杆藻（*Asterionella formosa*）。通过研究沉积物中硅藻集群和磷含量的传递函数可以发现，这种导致藻类水华暴发的营养物浓度在1920年后突然增加，和湖中娱乐项目的开发时间正好重合。梭罗独居瓦尔登湖时所享受的自然风光现在每年夏天都有数以千计的游客参观，而为了保证下一代仍能享用到这一标志性湖泊的文化价值和生态价值，我们必须通过持续和仔细的管理予以保护。

第三章

阳光与运动

> 当波浪在荡漾……湖水的蓝和反射的颜色混合,并随着波浪的形状、大小和方向变化。
>
> F. A. 福雷尔

弗索朗瓦·福雷尔在他关于日内瓦湖的著作中集中阐述了湖泊的物理环境:光线、温度、风、波浪、水流和混合过程。在他于莫尔日的房子附近的港口出入口,他注意到水以惊人的速度和规律在狭小的开口进出。然后他意识到这和整片湖水如跷跷板一般的摇摆运动有关。他和渔夫交谈并从中得知一些奇怪的事:他们布下的渔网经常在深水处被水流拽走,但是方向却和当时的风向相反。他意识到湖水不像充氧鱼缸里的水一样混合充分,而是由不同温度的水层组成,这种分层会随着季节变化而变化。

福雷尔对阳光和湖水之间的相互作用特别着迷。这种兴趣来自观察画家朋友弗朗索瓦·博西昂作画,后者将日内瓦湖五光十色的天空、云彩和水面捕捉到画布上。福雷尔观察到近岸湖水浊度的变化,由清澈到半透明再到浑浊不清。因此,他推测水的透明度是湖泊生态系统健康状况一个简单但有效的指标。今天的淡水科学家对所有这些特征的重要性都有清晰的认识,这些特征决定了湖泊的物理生境特点,并且对其化学性质、生物学性质和生态系统服务都有着强烈的影响。

清澈和浑浊的湖水

在福雷尔开始研究日内瓦湖后不久,他了解到一种测量水透明度的简易方法,成为第一个将这种方法应用到湖泊研究,并为其制定了标准流程的人。这种方法由一个担任教皇科学顾问的牧师彼得罗·安吉洛·西奇设想并提出,以测量地中海碧蓝清澈的海水。他在教宗国海军的"圣母无染原罪瞻礼号"上研究阳光和海洋的关系。他的方法从容又简单,即放下一个白色的盘并记录它不再可见时的深度。

福雷尔从西奇的基础上发展出一个标准化流程:取一个直径为20厘米的白盘,记下白盘沉至消失时的深度,再将白盘缓缓提起,记下重新出现时的深度,然后取两个深度的平均值作为"西奇深度"(即透明度)。西奇用过不同颜色和尺径的圆盘,其中一个直径为2.37米。福雷尔推荐使用直径20厘米的圆盘,除了便携,其效果也和稍大一点的35厘米版本没有不同。他同时在用一个有白色涂层的圆锌板和一个有白色釉质的瓷质餐盘,发现前者更加坚固,而后者的白色保存得更久。如今,在湖泊研究中最为常用的是直径为20或30厘米的金属圆盘,并涂有交替的黑白四象限图案以增加对比度。

透明度的值介于几厘米到几十米间，前者见于水华暴发的高污染水域，后者见于世界上最澄清的湖。世界上水域透明度最高纪录由南极洲的威德尔海保持，在那里，一个20厘米的西奇盘在沉至79米时仍然可见，这接近于纯水透明度的理论极限；而对湖水来说纪录保持者则为美国俄勒冈州的克雷特湖，一个直径为1米的西奇盘（尽管超出福雷尔的标准，但仍符合西奇牧师的要求）在水深44米处仍然可见。

在湖泊与海洋的研究中，阳光在水中的穿透程度可以通过水下光度计（潜水辐射计）更准确地测量。这些测量的结果始终表明，光强随深度的变化是一条锐曲线而非直线（图9）。这是因为阳光在水中向下传播的程度由光子被吸收或偏离光照路径的概率决定；假如光子在湖水中每米损失的概率是50%，那么水深1米处相比于水面的光强只有50%，到了2米则为25%，3米则为12.5%，依此类推。这种指数曲线意味着，除了那些最清澈的湖，只有上层水柱才有足够的能量满足植物和浮游生物（浮游植物）中的光合细胞的生存需要。

水下进行的光合作用或初级生产的深度极限被称为补偿深度。在这个深度，由光合作用固定的碳可以恰好抵消细胞呼吸中损失的碳，所以新生物量的总产量（净初级生产量）为0。这一深度的光强往往是水面光强1%的水平（图9）。在这一深度往上的区域都有通过光合作用产生的生物量，这一区域被称为真光带。更深、光线更少的地方被称为无光带，这里不可能进行光合生长，主要的生物活动限于摄食和分解。

用西奇盘测量透明度能够大致估算真光带的范围；通常1%光强的深度对应透明度的两倍。但是这个方法并不总是准

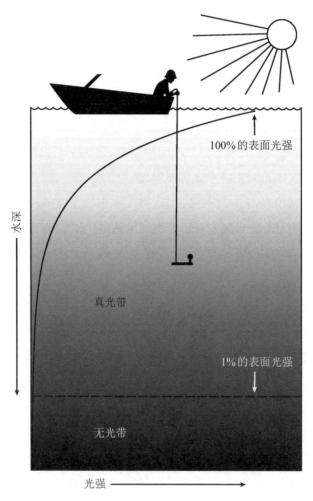

图9　用水下光度计测量得出的湖水透光度

29　确,因为光子流在水中传播,到达西奇盘并被反射重新回到水面这一过程中受两种因素影响:吸收(定义为a值)和偏转(又称散射,定义为b值)。这两个因素加起来决定了光线的衰减,定义为c值。a值和b值的权重取决于水中的颗粒和溶质,从而影响了透明度。研究自然水体的光学专家将透明度称为"表观特性",因为其数值取决于测量时光线的条件。如透明度在接近傍晚时所测得的数值偏小,这是因为太阳的角度较低;而在夜晚,透明度接近于无,哪怕月光格外明亮。另一方面,a值、b值和c值属于"内在特性",因为它们属于湖水的固有性质,不受测量时阳光条件的影响。

一般来说,水中的藻类颗粒越多,水吸收和散射光线的能力就越强。所以透明度是衡量湖泊富营养污染和藻类富集程度的指标。然而对于环北极带和亚马孙雨林地区的湖泊来说,由于森林环伺,入注的水富含来自森林腐殖土的褐色茶状物质。这些物质对光的吸收能力很强,掩盖了藻类的影响。另一极端则是水中含有大量反射性的矿物质悬浮颗粒。这些水有很高的b值,西奇盘反射出来的光子又被这些悬浮物散射,才回到我们的眼睛里,但它们仍能用于水中的光合作用。所以这种情况下透明度需要乘以一个系数(对矿物质颗粒丰富的水体而言这个系数至多为3)才能估计真光带的深度。

尽管有其局限性,西奇盘在湖泊研究和科学交流中依然是价值很高的工具。查尔斯·R. 古德曼是加州大学戴维斯分校的湖沼学教授。他从1960年代开始基于一系列参数长期测量和研究太浩湖,包括营养物、溶解氧、浮游动物生物量和光合作用。他发现在所有这些湖泊参数中,透明度随时间的下降对政策制定者来说是最容易理解和最有说服力的证据,可以促使他们制定严格的流域管控措施,以保护太浩湖湖水闻名遐迩的澄澈和蔚蓝。西奇盘在当下仍是湖泊研究中的常用工具(通常和其他潜水式光学仪器联用)。由于其价格低廉,操作简便,在世界上许多地方的各类公众外展服务活动和公民监测计划中,都能够找到西奇盘的身影。如北美五大湖管理协会(NALMS)每年都会组织数以百计的湖边居民和来自美国与加拿大各地的游客,进行"西奇盘下水"的活动。

水的颜色

福雷尔对不同湖泊,甚至同一湖泊中不同区域呈现的不同颜色特别感兴趣。他发展了一套液体色标(现在已有手机应用程序),将水按照不同的颜色归类,然后进行实验,试图探究这些差异的原因。他无法想象的是,现在各种强大的测量方法都已将水的颜色纳入测量范围,以追踪湖泊、河流、河口和海洋的水质以及其他特性。有越来越多的光学仪器面世,从可以沉入湖中测量不同光谱带(包括紫外线)的断面仪,到可以全年自动测量的锚泊系统,以及从太空持续监测湖水颜色变化的卫星。

根据其中溶解物和悬浮物的不同,湖水有不同的颜色、色调和亮度。最纯净的湖水是深蓝色的,因为水分子能够吸收绿光和红光,后者的吸收程度更高;剩余的蓝光光子则被散射至各个方向,绝大部分都向下,但也有少部分回到我们的眼睛里。有的湖泊,如南极洲的万达湖和俄勒冈州的克雷特湖,其湖水如墨水一般呈深蓝色,仿佛将手放进去就会被染成靛蓝色一样。

水中的藻类一般会引起水色变绿和水质浑浊,因为这些藻类细胞和群落含有叶绿素以及其他捕捉光的分子,能够强烈吸收蓝光和红光,但不会吸收绿光。然而总有特殊情况。有害藻类水华如果由蓝藻引起则湖水为蓝绿色(青色),这是因为它们除了叶绿素以外还有蓝色的藻胆色素蛋白。在英国,一些乡村里的池塘被称为"红绿灯池",因为它们会在一天之内从绿色变成红色。这是由裸藻门的浮游藻类造成的,它们除了含有光合色素,还有红色颗粒。在光线昏暗或完全黑暗的条件下,这些红色颗粒藏在细胞内部,而亮绿色的叶绿体则完全暴露于外部环境中。然而,当阳光明媚时,红色颗粒会迁移至外部遮蔽叶绿体,以抵御太阳光辐射造成的伤害。某些水产养殖的鱼塘管理员会惊诧地发现,他们那些富含裸藻门藻类的鱼塘会在烈日下突然变成亮红色。其他淡水藻类也能引起水质变红,如引起赤潮的辐尾藻,如含有藻红蛋白的浮丝藻属的蓝细菌,又如遍布世界各地花园中鸟浴盆里的血红色的红球藻。

　　福雷尔在日内瓦湖入湖口和近岸的水边所观察到的黄色来源于流域冲刷下来的溶解有机质。具体来说这些溶解物是一系列高分子量的有机混合物,被称为腐殖酸。这些茶状物质由树叶在土壤中降解所产生,然后被冲刷到湖中。这些物质以前被称为 gelbstoff,从德语直译为英语大意为"黄色物质"。在1895年,福雷尔提到:"湖水中的这些有机物的性质是什么?这个问题还没有得到充分的研究。"在这点上,他是完全正确的,因为在一百多年后的今天,这一问题依然是湖泊和海洋科学中的研究热点。现在这些金色物质被称为"有色溶解有机质",简称 CDOM。由于我们对这些复杂的混合物的化学本质依然只有有限的了解,这一现代词汇仍有些含糊不清。

　　CDOM 一个有趣的特点在于它可以强烈吸收蓝光,并对紫外线的吸收能力更强。因此,CDOM 成为河流和湖泊天然的防晒屏障,保护水生生物群免受有害紫外线的灼伤。CDOM 对湖泊颜色的影响取决于其浓度。在最高浓度下,太阳光谱内所有波长都被吸收，湖水也被染成浓缩咖啡一般的黑色。而在较低也是更常见的浓度下，CDOM 吸收蓝光和蓝绿光，使湖水呈金棕色。在最低浓度下，CDOM 吸收蓝光，同时水分子吸收黄光到红光的部分，可见部分只剩下绿光。福雷尔通过一项实验证明了颜色和 CDOM 浓度的关系。他首先把富含 CDOM 的棕色沼泽水过滤，然后用日内瓦湖清澈的水将其稀释，再装入一根玻璃底的长管，以便他从底部向天空观察。这时候的水澄清透明，并呈柠檬绿色，如他在日内瓦湖沿岸区入流水和湖水混合处所看到的一样。

神秘的水

水在我们的生活中如此常见，以至于我们对它熟视无睹。我们也不会在打开水龙头或喝一口饮料时把水看作一种化合物。然而水是一种拥有各种奇怪性质的化合物，而且其中一些尚未得到完整的解释。这些性质对湖泊的物理、化学属性以及生活在其中的水生生物有着巨大的影响。

这些奇怪性质的核心是水分子本身，以及它倾向于以不停变化的有着不同大小和复杂度的簇聚集在一起的特性。每个氢原子和氧原子之间都共享一个电子，形成一个共价键，构成水分子。但因为氢原子只有一个带正电的质子，是氧原子的 1/8，因此，在这个关系中，氧原子更像是一个老大哥，将带有负电荷的共享电子云稍微拉向自己这边。这导致氧原子带轻微的负电，而两个氢原子带轻微的正电。异性相吸，因此水分子粘在一起，氧原子和其他水分子中的氢原子因为静电力结合成为氢键（图10）。每个水分子最多能够和其他四个水分子形成氢键。尽管仍有争议，液态水中大部分的水分子都是动态地连接在一起，以其中一个氧原子为中心，构成金字塔形的结构（四面体）。

水的另外一个奇怪性质是其独特的密度与温度的关系。通常来说，相比液体形式，物质的固体形式更紧密，密度也更高。然而冰却完全相反，因此可以浮在水面上。这是因为在冰里面，所有的水分子都以氢键和其他四个水分子相连，这样的晶体阵列中分子间距达到最大值。一旦冰融化，液态水分子不再有全氢键相连的条件，这种膨胀结构随之消失，分子变得更致密，导致密度上升。结构的松弛程度会一直随温度的上升而加剧，直至4℃左右（准确来说是3.984℃，在标准大气压的条件下），此时水的密度最大。而随着温度进一步上升，水分子动能的增加和运动的加剧导致分子之间的距离上升（尽管无法和冰相比），而密度下降（图10）。

为什么这种密度-温度的关系那么重要呢？对加拿大人和

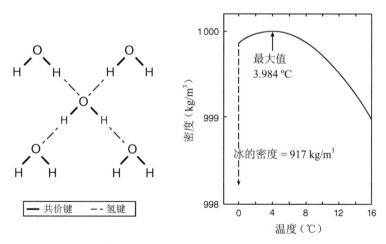

图10 水的氢键和其密度与温度的异常关系

其他北方人来说，这意味着我们对坚硬的固态水会浮在表面上有着十足的认识，并希望它足以支撑我们在冻结的液体上活动，能够让我们冬天在结冰的湖上滑雪、穿雪鞋徒步和开雪地摩托。在全世界范围内更广泛的意义上，它还意味着暖水总是在冷水之上。所以在夏天湖泊温度上升时，一层暖水会浮在底层更致密的冷水上。而这两层水的交界处被称为温跃层，其上下层相应被称为湖上层和湖下层。这两层水除温度不同之外，其化学性质和生物学性质也不同。

湖泊的季节与混合

从冰层覆盖的寒冬到暖水浮在冷水上的炎夏，湖泊的分层程度（或称分层现象）会随着季节而大幅变化。任一时间内的不同水层在物理和化学性质上都有显著的不同。举例来说，魁北克市的圣查尔斯湖是我们的饮水源。在晚夏的时候，其温跃层大约在湖面7～10米下（图11）。在这段深度中，除了温度大幅下降，氧气含量也急剧下降。湖上层因为可以和上层大气交换氧气，所以氧气含量接近饱和。温跃层充当了阻挡氧气交换的角色，而底层湖水则完全没有这种维持生命的气体。因此，晚夏至早秋的圣查尔斯湖湖下层，对鳟鱼这类需要充足氧气的鱼来说毫无吸引力。

随着湖上层在秋季温度下降，湖面和湖底的温差逐渐变小，阻挡上下层湖水混合的密度层也随之消失。另外也是最重要的，由于表层湖水温度下降、密度增加和下沉，形成对流循环，它与风致混合协同，最终导致整个水柱的混合，这被称为湖水对流。此时底层湖水得以补充来自大气层的氧气（图11）。而由

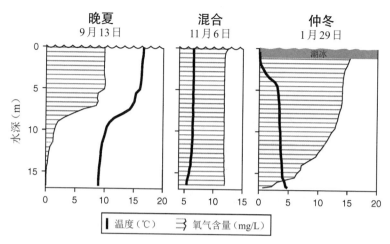

图11 魁北克市的蓄水库圣查尔斯湖的季节性变化。底部刻度为温度[粗线,单位为摄氏度(℃)]与氧气含量[阴影区域,单位为毫克每升(mg/L)]共用

于所有气体(包括氧气)的溶解度在冷水中更高,在混合期结束后,湖水的含氧量上升至比夏天温暖的湖上层更高的水平。

在气候更加温和的温带,湖在冬天可能不足以冷至结冰。这些湖被称为单循环湖,因为它们只有一次垂直混合时期,即使这个时期很长。随着湖水从秋天到冬天冷却和混合,氧气一直得以补充。世界上的许多湖都是单循环湖,如琵琶湖、日内瓦湖、太浩湖、的的喀喀湖、陶波湖、马焦雷湖以及英格兰湖区的一众湖泊。

相比于水中化学和生物过程对氧气的需求,氧气在水中甚至在冷水中的溶解度都不算高。在所有的湖中,这种至关重要的气体的收支总平衡并不稳定。这对冬天覆有冰盖的北方温带湖来说尤为明显,如圣查尔斯湖(图11)。尽管经过冬天前的冷却和混合,这些湖泊的氧气充沛,冬天的覆冰和积雪却不但切断了湖水与上层大气的氧气交换，也屏蔽了光合作用需要的光线，阻止氧气的产生。同时，分解作用也在持续消耗氧气，特别是在沉积物中，最终导致底层水完全没有氧气（即缺氧）。在整个冬天，湖泊高度分层，此时冷水（密度较低）浮在稍微暖一点的水上。这被称为逆向分层，因为温度曲线是相反的（冷水浮在暖水上），但水的密度随深度增加而增加，以达到引力平衡。

在春季，湖冰融化后，加拿大人将他们的雪地靴收起，同时湖上层的湖水逐渐变暖至下层的温度。温差消失后，密度差也消失了。在风的作用下，湖水自上而下混合和补充氧气。这种湖因此有第二个混合季节，所以被称为二次循环湖，全年共有两个全水柱混合时期：秋季和春季。然而和秋季不同的是，春季的混合会随着温度的季节性变化迅速减弱：一旦表层水升至4℃以上，其密度便比春寒时更小，因而浮于湖面，成为一层持续变暖的湖水，阻挡进一步的混合过程。因此，二次循环湖的春季混合过程会很短，有时和秋季持续很久的混合相比可以忽略不计。

湖面活跃的波浪

当风轻轻拂过一面湖时，会让水面产生涟漪。这些小波浪会被风的阻力带起来，由于水分子之间的氢键将其拉回湖中，波浪又跌入波谷。这些小波浪被称为毛细波，强调这种恢复力来自毛细级别的水分子作用力（也就是表面张力）。这种小波浪的最大波长是1.73厘米，周期不超过1秒。随着风力越来越强劲，这些波浪会被拉得更高，此时的恢复力则由重力主导。当风速高于每小时25～30公里时，或当波浪进入更浅的近岸地区时，顶层的波浪运动得更快，拉长了底部的波浪并破碎产生白浪，这使得表面水得到强烈的混合和充氧。重力波浪的最高纪录在北美五大湖，波幅（即波谷到波峰间的距离①）高达8米。但是因为湖泊上的风达不到海洋上的规模，大部分的波浪都不足50厘米高。

　　乍一想，表面重力波浪有着大量的能量，似乎足以混合湖水。波浪的确能引起底层水运动，尤其是一系列的环形运动，其直径随深度指数性下降，但不足以充分混合水柱。这种波浪运动在湖泊沿岸区（近岸区域）能够使细小的沉积物重新悬浮，这说明细小的沉积物只能在远离湖岸的地方积累。在深水区，这种波浪的影响无法穿透。沉积物重新悬浮和积累的临界深度被称为"泥沙沉积边界深度"，这一深度取决于风暴事件引发的波浪波高以及沿岸区湖底的坡度。然而，湖泊更充分的混合不依赖于这些活动的波浪，而依赖那些对大多数湖泊游客来说不太明显的、更缓慢的波浪。

湖面上下缓慢的波浪

　　福雷尔在他的自传中指出，他最喜欢的一个研究课题是湖泊如同钟摆一般缓慢的摇摆运动。这种现象在日内瓦湖很知名，以至于当地人用瑞士法语方言给这种现象取了一个名字——*seiche*，意为"假潮"。这个名字现在逐渐被世界各地的科学家使用，以描述湖泊这种无处不在的现象。假潮最明显的特征是湖水水位的改变，起伏周期为数分钟至数小时，尤其是在

---

① 波幅的实际含义为波偏离水平中线的垂直距离，应为波谷到相邻波峰间垂直距离的一半。原文应有误。——编注

靠近岸边的地方。

福雷尔通过构想和安装各类水位连续记录仪,包括一个便携式记录仪,对日内瓦湖和其他湖泊的假潮做了仔细的观察。他初步尝试推导一个假潮的数学理论,但对这个尝试感到沮丧,并在自传中后悔地承认大学时放弃了修一门有用的微积分课程,只因为那个老师特别没有启发性(这也是给所有教授上的一课)。

但福雷尔有其他的热情和天赋,包括和欧洲及世界上其他地方的科学家建立良好的关系。他联系了当时一位著名的物理学家,即后来被册封为开尔文勋爵的威廉·汤姆森。汤姆森帮福雷尔把一个笨拙的等式简化成一个优雅的形式:

$$P = 2L/\sqrt{(gh)}$$

其中 P 是湖水水位起伏的周期,L 是湖泊的长度,g 是引力常数,h 是平均湖深。福雷尔正确地猜想到假潮是横跨整个湖泊长度的驻波,而且还伴有振幅更小的次级波。

那么这些振动的根源是什么?福雷尔得到正确的结论:假潮来源于风,它持续地将湖水吹向和推向湖的一端(图12)。这种"设定"好的条件让湖水水位在下风向一端高,上风向一端低。但这种水位并不稳定,一旦风停下来,湖水会向后晃动并在另一端升得过高。这种跷跷板运动会持续至湖面初始错位的势能最终被耗尽为止,就像钟摆最后停止下来一样。

假如知道假潮对湖泊最大的影响在湖面以下,福雷尔应该会感到很惊讶。这种影响可以达到温跃层的深度,那里的波动和混合会影响氧气和养分的运输。他的研究表明温跃层深度可

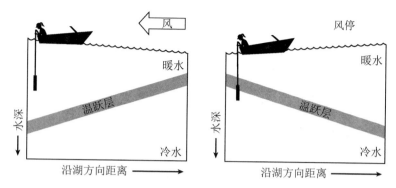

图12 表面假潮由风将水吹向湖泊的一端而引起,其引起的内部假潮可以作为温跃层的一种振荡来检测

以在短时间内变化,但是他当时没有把这种变化和表面水位的变化联系起来。直到欧内斯特·M. 韦德伯恩和其他科学家做了苏格兰湖泊的经典研究后,"内部波"或"内部假潮"的本质才被揭示出来。当湖水被风堆积在湖的一端时,覆盖在湖上层的质量更大的水会把温跃层向下推。而当风停下来,温跃层会再次上升并且过高,然后持续地振荡直至初始势能耗尽。

图12给出了温跃层的风致运动的一个总体概念,但是需要进一步的说明。假潮在垂直方向的规模被夸大了,而且没有体现假潮的一个重要特点,即内部假潮比湖面假潮更慢且周期更长。例如在日内瓦湖,沿湖泊主轴的湖面假潮周期约为74分钟,而主要的内部假潮(伴随其他更高频的更小的波动)周期则长达3天,远比风诱导起势而致的湖面假潮散去所需的时间长。图中的淡水科学家需要固定地守在湖面几个小时到几天,才能正确合适地观察到缓慢的内部波运动。然后她会从所得的水下热剖面注意到,任意深度的湖水温度都在逐渐上下波动,而在温跃层的内部和附近变化最大。

由于多种原因,淡水科学家对内部波保持着浓厚的兴趣。首先是波幅的问题:这些内部波的波幅可能是巨大的。表面假潮是空气与水界面的错位,由于两种流体的密度差,哪怕是下风端水位(和相应的势能)小幅度的上升也需要消耗大量的风能。因此,湖面假潮的幅度通常较小,介于几厘米到几十厘米之间。当然也有例外,如伊利湖上风暴引起的5米高的假潮。另一方面,对于内部波来说,上层水和深层水之间的密度差异很小,特定水面错位带来的同样势能在作用时产生的温跃层错位很大。在深水湖如太浩湖中内部波的初始阶段,底层水在垂直方向上的错位可能高达100米,引起在上风端的湖水上涌。这种上涌将营养物带到真光带,刺激了藻类的生长。

湖泊和海洋中的水运动受地球自转偏向力的影响,内部波也不例外。随着波在温跃层的振荡,它们在南半球的湖中会偏向左边,在北半球则向右。这种所谓的"科里奥利效应"对水运动的影响相对较弱,但在中型到大型湖泊中可观察到它以两种方式影响着内部波。首先,这种效应会将波困在湖边,并引导其以逆时针(北半球)或顺时针(南半球)方向运动。开尔文勋爵首次发现这一存在于大型湖泊、大气和海洋中的现象,并予以正式定义。"开尔文波"这个名字同样很好地引用了这位帮助福雷尔发展假潮理论的物理学家的名字(也许开尔文勋爵在讨论的过程中启发了自己关于波的想法),因为这种波对许多湖都有重要影响。例如在日本的琵琶湖,开尔文波会被台风季节的强风激活,有时能到达湖面,把冰冷且营养丰富的水带到表层并绕湖做环形运动。

受科里奥利效应影响的第二种内部波发生在远离湖岸的

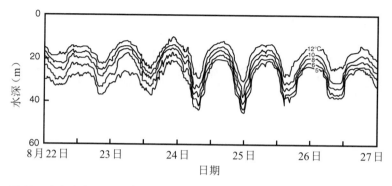

图 13　安大略湖温跃层的庞加莱波

湖泊主体处,被称为"庞加莱波",得名于杰出的法国数学家和理论物理学家亨利·庞加莱。这两种波的波幅都很高。例如图 13 显示了位于加美交界的安大略湖一个固定采样站测得的五天温度数据。观察所得的波的性质令人惊讶,波谷到波峰的距离高达 25 米。庞加莱波比近岸的开尔文波的周期短得多,但仍比表面假潮长得多。在安大略湖,庞加莱波的周期是 16 小时(图 13),而开尔文波为 10 天,沿湖泊主轴的表面假潮为 5 小时。

　　湖泊生物学家和生物地球化学家都对内部波特别感兴趣,因为这些内部波对分层湖水的混合有着决定性的作用。如图 11 所示,这些分层湖水在温度、氧气含量及其他性质上都极为不同。这些由波引起的湖水水平运动延伸至湖床:随着水在湖盆中前后滑动,一种振荡的湍流在沉积物上穿梭,将颗粒向上带至一个叫作"水底边界层"的区域并在那里悬浮。这种水的流动和混合将氧气带到沉积物中,并加速其他化学物如氮和磷在湖床与湖水之间的交换。与浮游生物特别相关的是,在初始和随后的振荡中,倾斜的温跃层会把深层水带到表面,并让这些深层

图14 横跨温跃层的波状运动。这些波动由前进波发展而来,对湖上层和湖下层的物质进行滚动搅拌、破碎和混合

水参与表面混合。这把营养物从深水区挟带到真光带,并引起水平流动,帮助湖水混合均匀。

然后还有"波状运动"(图14)。内部波在温跃层上下引起水的反向运动,这带来了阻力和压力梯度,随之引起短暂的沿着温跃层传送的前进波,叠加在内部驻波上。这些高频波一般周期只有大约100秒,波长在10～15米之间,波幅在0.05～2米之间。最重要的是,这些波会卷起并破碎,就像表面的波浪一般,在温跃层的屏障上短暂地打开了一个窗口,以进行热量、营养物、氧气和颗粒的交换。这些层状流体中的波状运动效应被称为"开尔文-亥姆霍兹不稳定性"(依然得名于开尔文勋爵,但这次联同著名的德国物理学家赫尔曼·冯·亥姆霍兹)。天空中的云有时也会出现这种不稳定性产生的涡旋,这是由于冷暖空气的混合所致。波状运动的出现取决于温跃层中的速度梯度。这种现象可能在湖边出现,引起营养物供应和初级生产活动的增加;或在湖心出现,促使整个湖泊内的藻类生长。

湖中的水流

与湖面假潮和内部假潮相关的振荡流动只是湖水各种眼花缭乱的大量运动的一部分而已。福雷尔指出,在最大的维度上,即在整个湖泊的层面,从入流河到出流口必定会有一个净流量。而且他建议从地表景观的角度考虑,可以把湖泊视作扩大的河流。当然,这种如河流一般的流动不断地被水的风致运动所干扰。当风刮过水面时,会连带水产生一个下风流,这必然会被深处水的回流所平衡。它解释了为什么日内瓦湖的渔民会发现,撒在深水处的渔网会被拉向一个和盛行风相反的方向。

在大型湖泊中,随着湖水从湖的一侧转移到另一侧,地球的自转有足够的时间施加其微弱的作用。因此湖面的水不再沿直线前进,而是被引导为两个或更多个环形运动或者涡流。这些涡流能把湖边的水快速卷入湖心,反之亦然。所以涡流有着巨大的影响,可以将水及其携带的物质快速地从一处转移至另一

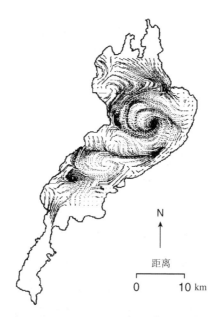

图15 日本琵琶湖的涡流,数据由科研船只"发现号"配备的声学多普勒流速剖面仪测得

处，这些物质包括污染物甚至有毒藻类。这会形成惊人的水流图案。例如，日本琵琶湖拥有被认为是世界上最美的涡流。与科里奥利效应无关的是，即使在小一点的湖泊里，风致水流和湖岸线的相互作用也会导致水在湖边做单个的环形流动。

在更小的层面上，刮过湖面的风会引起下风处水的螺旋运动，被称为"朗缪尔环流"。这个现象由杰出的美国科学家欧文·朗缪尔在马尾藻海首次发现。在大洋的这片海域，相邻但方向相反的螺旋水运动把海面的浮物聚拢到一处。这些浮物主要是马尾藻属的海藻，会累积形成长条的平行线。湖泊中常能观察到这些螺旋运动，它们沿着风的方向，把泡沫（在有白浪的日子里）或光滑的油状物（在风不大的日子里）聚拢成规则间隔的直线，并和风向平行。

在这个简单讨论水运动的章节还必须提及密度流。它们在世界上绝大多数湖泊中都发挥着重要的作用，包括日内瓦湖。如同分层现象，这种现象由水的密度-温度关系引起。寒冷的河水进入温暖的湖泊时，会因密度更大而沉入底部，而且会继续向前流相当长的距离。在日内瓦湖，来自罗讷河上游富含沉积物的寒冷河水进入湖中后，会立即潜至湖床并沿后者前行数公里，切割出一条被称为罗讷谷的沟谷。利用多波束回声测深仪对沟谷的测量表明：在某些地方，沉积物以每年几米的速度被侵蚀；而在其他地方，河水挟带的沉积物则累积在底部。通过这种方式，这个水下沟谷成为湖底一个不断变化的蜿蜒峡谷，也成为冰川入流水的水渠和蚀刻通道。密度流对近岸和远岸水的交换起着巨大作用，对初级生产力、深水充氧和污染物的分散也有潜在的影响。

第四章

生命支持系统

> 据我所知,没有严格的分析可以表明,有任何湖水是完全不含微生物的……
>
> F. A. 福雷尔

弗朗索瓦·福雷尔认为微生物广泛存在,这在大体方向上是正确的。但是,他却没能猜测到,支撑起自然水体生态的微生物有着惊人的多样性和丰富度。即使从最清澈的湖中舀起的一杯水,也蕴含着一个看不见但又拥挤的世界:在这之中也许悬浮着十万个光合细胞,一千万个细菌细胞,和一亿个"野生病毒",而这些肉眼都看不见。在很长一段时间里,淡水生态学家都对这些共栖在均匀的表层湖水中的不同种藻类细胞感到困惑。G. 伊夫林·哈钦森是最知名的湖沼学家之一,他将这种共栖现象称为"浮游生物悖论":为什么一种生物不会通过竞争以获得有限资源,从而将其他的生物赶尽杀绝?随着新的分子生物学和生物化学技术的发展,微生物那令人叹为观止的多样性更为显著了。如同我们现在知道"人体微生物组",即生活在我们体内和体表的微生物群,会影响我们的健康状态,"水生微生物组"也对湖泊生态系统的健康运作和系统对环境变化的反应起着至关重要的作用。

太阳能经济

地球上几乎所有的生态系统都依赖太阳的能量输出,要么是直接从当前进行的光合作用中获取,要么就是间接地通过过去光合作用合成所累积的生物质,以及之后微生物和动物对它的消耗来获取。旧植物材料的循环使用对湖泊来说尤为重要。理解其重要性的一个方法是测量二氧化碳浓度,它是分解作用在表层水的最终产物。这一数值往往高于有时甚至远远高于和上方大气达到溶解平衡时的值。这说明许多湖泊都是二氧化碳的净生产者,将这种温室气体排放至大气中。这是怎么回事呢?

要找到这个问题的答案,我们要把目光放到充水的湖盆以外去。湖泊不是封闭独立的微观系统,相反它们处在地表景观当中,并和周遭环境紧密相连。湖中的有机物由浮游植物产生,它是一种悬浮在水中的光合细胞,能够在湖中固定二氧化碳、产生氧气,并为水中的食物网底层提供生物质。在湖边阳光能够到达底部的区域,附生藻类(周丛生物)和水下植物(水生植物)也能在此生长并进行光合作用。但除此之外,湖泊还是其流域地表径流的下游接收者,除了接收水以外,还有随溪流、河流、地下水和地表水冲刷下来的土壤有机碳以及植物补贴。

从流域进入湖泊的有机碳被称为"异源碳",意思这些有机碳是从外部进入湖泊的。而且因为这种有机碳是由之前陆上的植物生产的,所以年份相对较为古老。与之相反的是,沿岸区的植物群落和浮游植物通过最新光合作用合成的有机碳则要年轻许多,它们供应着微生物和食物网。这些有机碳被称为"自源碳",意味着这些有机碳是在湖泊内部合成的。这些年轻的溶解有机物大部分都是小分子,会很快被湖泊中的微生物消耗掉,是碳和能量的首选来源。另一方面,大多数异源碳由腐殖酸和富啡酸组成,后者从地表植物材料衍生而来。这些茶色酸都是含有有机碳环的大型聚合物(图16),可以抵御多数细菌的分解,不过还是会被某些真菌分解掉。

　　人们一度以为,大多数进入湖泊的溶解有机质(DOM),尤其是带颜色的部分,在化学上都不活泼。这些有机物会在通过

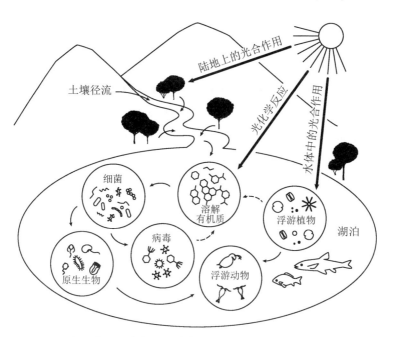

图16　从阳光到多种微生物和水生食物网

湖泊时保持不变,并最终从出流口离开。但是,许多实验和实地观察都显示这些有色物质会被阳光部分分解。这些光化学反应会产生二氧化碳,同时将部分有机高分子降解为有机小分子。这些小分子又会被细菌利用并分解为二氧化碳。这种由阳光驱动的化学反应从河流开始进行,并在湖泊表面水层继续。其他在土壤中进行的化学和微生物反应也会分解有机物质,并将二氧化碳释放到径流和地下水中,进一步导致了湖水中的浓度变高,并被排放到大气中。在充满藻类的"富营养化"湖中,充足的光合作用可以降低二氧化碳的浓度,直到湖水与空气的平衡值以下,从而引发这种气体从大气向湖泊的逆向流动。

失衡的氧气

尽管湖面上大气中的氧气似乎取之不尽,湖泊中氧气的收支平衡却并不是稳定的。当氧气几近甚至完全被消耗时,这一氧气平衡会产生巨大的赤字以至于湖泊出现氧气破产。氧气浓度在 2 mg/L 以下的水体被称为"低氧",大多数鱼类都会避开这种环境。而水中完全没有氧气的时候被称为"无氧",大多是适应力极强的特殊微生物的领域。例如,亚马孙温暖的水域在夜间会转为无氧的状态,鱼类必须迁移到氧气充足的浅水区或到亚马孙河里,而捕食种如鲶鱼就在这些充氧的水域中以逸待劳。在许多温带湖泊中,春季和秋季的混合期对湖水从上层大气中再次获取氧气来说很重要。然而在夏季,温跃层会大幅降低氧气从大气到深水的转移速度;而在寒冷的季节,湖水覆冰又是充氧的另一层屏障。在夏冬两季,湖水在之前春秋两季所补充混合的氧气可能会被迅速消耗,变成无氧的状态。

湖泊长期处于无氧危机的部分原因在于氧气在水中的溶解度很低，导致水中的氧气储存有限。氧气在空气中的浓度为209 ml/L（体积百分比为20.9%），但在冷水中的溶解度只有9 ml/L（体积百分比为0.9%）。随着温度的升高，氧气在水中越显稀缺（从4℃至30℃，氧气溶解度会下降43%）。而由于在暖水中细菌的分解更快，所以对于呼吸的氧气需求就更高，因此加剧了这种稀缺。此外，微生物对这种珍稀资源的需求本身就很高，这是因为湖泊是整个地表景观中分解作用非常活跃的地方，充斥着大量的耗氧细菌，从流域进入以及自湖中产生的有机碳是这些细菌的燃料（图16）。

识别不可见的世界

1904年，福雷尔在他关于日内瓦湖微生物学的描述中指出，尽管细菌无所不在，但是人们不需要害怕它们："绝大多数这种微小的生物都是无害的。"也许它们是无害的，但是考虑到它们巨大的种群数量和物种数量，以及在功能上的多样性，它们绝不是微不足道的。直到最近，要识别现存的细菌和它们在湖泊生态系统中的特定功能都是不可能的。因为大多数细菌都不能被带入培养皿中，而且也无法仅仅通过在显微镜下观察就进行识别和区分。然而到了今天，基于核酸测序的方法让人们无须经过问题重重的培养就可以分析湖泊的样品，尤其是针对DNA（脱氧核糖核酸）的测序可以用来检测种群中的基因，而针对RNA（核糖核酸）的测序可以检测基因在生产蛋白质过程中的表达。这些分析能够揭示湖泊种群中非常多样的活跃微生物和微生物功能，这些元素共同组成了湖泊的微生物群系。人们对许多微生物属和微生物物种仍知之甚少，同时每年又会发现新的微生物和微生物功能。

微生物群系由四个微生物组组成，而这四个组分各自又有着巨大的物种多样性和功能多样性。规模最小、数量最丰富，也许是最多样的当属病毒。这些微缩版的寄生物会攻击细胞并且对其进行重新编码，以生产更多的病毒颗粒。这些颗粒一般介于20～200纳米（毫米的一百万分之一）之间。每种细胞类微生物都有自己的一组病原病毒，而由于细菌是微生物群系中最丰富的细胞，大多数在水中自然发生的"野生病毒"都是细菌寄生物，被称为"噬菌体"。但是其他的病毒会攻击微生物食物网中的其他组分，这可能会引起这一食物网季节性的变化，而变化的程度尚未得到很好的认识。一组被称为"巨型病毒"的病毒（拟菌病毒及其亲属）吸引了许多关注，因为它们有着大尺寸（超过250纳米）和能够寄生于变形虫及藻类的特点。还有其他一些攻击水生动物的病毒，如寄生在鱼类中的"传染性造血组织坏死病毒"（IHNV），这种病毒会使鳟鱼和鲑鱼感染，并导致渔场中的鱼大量死亡。

一旦病毒的后代在繁殖阶段完成组装，会引导宿主细胞进行"溶胞"（即爆裂），并从宿主细胞中释放出来。这一过程会将细胞的其他物质释放到水中。这些物质会作为基质被逃过病毒攻击的细菌所吸收利用。但这一胜利不会持续太久，因为它们可能又会被病毒感染，或被原生生物吃掉，后者又会被浮游动物吃掉（图16）。这种有机碳从一个细菌通过溶胞转移到另一个细菌，并最终到原生生物的分流过程被称为"病毒分流"。在一些湖泊和海洋环境中这可能占据了总碳流的10%。病毒还有另外一个角色,即在宿主之间转移DNA片段,这会修改宿主细胞的基因或授予它新的基因,并将这种改变一代代传下去,前提是宿主在这场旷日持久的微生物战争中得以幸存。

微生物群系的第二组分就是细菌本身,细菌"生命之树"的许多分支(门)都能在湖泊中找到。认识这种丰富性和多样性的其中一个方法就是用荧光染料将湖水样本中的细胞染色,过滤后放在一层薄膜上,并用荧光显微镜观察这层薄膜。你会看到样本像银河一样群星闪耀,发着荧光的细胞有着不同的尺寸和形状,大多数都是球形的(即球形菌),也有椭圆形的(即棒状菌)、螺旋形的、肾形的和丝状的。在这个微生物星系中,大部分细胞都非常小,直径介于200～400纳米,以至于福雷尔无法用他的标准显微镜看到。这些所谓的"超微细菌"的优势就是比表面积很大,这在湖泊环境中可以增加它们碰撞并吸收(通过细胞外膜上一些特定的"传输"蛋白质)有机分子和营养物的机会,而这些化合物在湖水中的浓度并不高。

最普遍的细菌门类是变形菌门,湖水中有三个最显著的亚门:α-变形菌,β-变形菌和γ-变形菌。其中以β-变形菌数量最多,占据了总浮游细胞数的70%。β-变形菌当中包括了一个恰如其名的属 *Limnohabitans*[①]。这种细菌似乎能够利用浮游植物释放的有机物质快速生长,从而抵抗捕食和病毒攻击的双重压力。另一种在全世界湖泊中都常见的β-变形菌——多核杆菌属,能够有效利用复杂的有机物质,包括从流域来的腐殖酸分解产物。一类在氮循环中起重要作用的β-变形菌叫亚硝化单胞

---

① 直译为"湖沼生物"。——编注

菌属，它们能通过消耗大量氧气将铵根离子（NH_4^+）氧化成亚硝酸盐。

γ-变形菌大部分分布在海水中，但在湖水中也有两个值得提及的科。甲基球菌科包括几个以甲烷为碳源和能源的属，如甲烷球菌属和甲基杆菌属，常见于能够产生甲烷的无氧环境，如湖泊沉积物的表层。另一个科是肠杆菌科，包括最臭名昭著的大肠杆菌（*Escherichia coli*）。这一细菌由奥地利儿科医生特奥多尔·埃舍里希从婴儿的粪便中分离得来，因此以他的名字命名。大部分大肠杆菌都不致病（尽管有一些危险的例外），但它们仍是监测湖泊的饮用水源和游泳区是否受到人类大便污染的指标。除了大肠杆菌中的感染性菌株，这种污染可能还包括其他水传播疾病的病原体，可能引起诸如霍乱、肝炎、伤寒和胃肠道疾病等病症。

湖泊中的大多数细菌都是分解者，能将有机物转为矿物质，如二氧化碳、铵根离子、磷酸根离子和硫化氢（H_2S）。除了这一重要的分解回收功能，一些细菌还擅长将无机物分子转换为能量来源（如硝化细菌），另一些则依赖阳光生存。后者在荧光显微镜下呈明亮的橙红色球体，在染色细胞组成的星系中尤为明显。这些是"微型蓝细菌"，虽然比分解者大，尺寸还是很小，在2微米以下。它们或许不曾被福雷尔用标准显微镜观察到，但也许是日内瓦湖，乃至绝大多数湖泊和海洋中数量最为丰富的光合细胞。它们在荧光显微镜下之所以呈橙红色，是因为除了叶绿素以外，它们还有蓝色和红色的蛋白质色素能够吸收光线并发出荧光。

古核生物是微生物群系的第三个组分。古核生物有着细菌的部分特征，如尺寸小、形状各异和"原核结构"，即没有细胞核和其他更高级的真核细胞（包括我们人类自己的细胞）所有的细胞结构。然而这种简单的结构掩饰了它们在基因和生物化学性质上不寻常的特征，这些特征使得它们和细菌以及真核生物如此不同，乃至被微生物学家分类为生命系统的第三个域。其中一些古核生物在自然水域中发挥着重要的生化功能，如生产甲烷及氧化铵根离子。

湖泊微生物群系中最后但绝对同样重要的一个组分是真核微生物。这些微生物有着更加复杂的细胞核结构，也叫"原生生物"，包括在历史上根据功能分出的两个大类：光合原生生物或藻类，它们通过捕获阳光进行光合作用并利用二氧化碳作为碳源；以及无色原生生物，它们以从湖水中吸收或细菌中提取得来的有机分子作为能源。即使是最健康的藻类在生长和繁殖的过程中也会释放一些光合作用的产物到湖水中，而在被病毒胀破或被浮游动物破坏时则会放出更多有机物质。假如这些有机物质没有被细菌摄取并随后被原生生物捕获的话，它们就会永远消失在食物网以外。这一碳恢复过程被称为"微生物环"，其中的一部分能量和碳会随浮游动物上行到食物网的上端，并最终为鱼类所用（图16）。

不久之前，生物学家仍以碳和能量的来源为依据，对生物界做出明确的划分：无机对应有机，光合作用对应捕食行为，植物对应动物。但是原生生物并不严格遵守这种划分，因为它们当中的许多个体能够在动物和植物的生活模式间转换。这种依赖于混合能量来源的生物被称为"混合营养生物"，而且在大多数湖水中都很常见。这种生存策略让它们既能通过光合色素获得太阳能，又能利用环境中已有的有机物所含的化学能。它们对细菌的捕食特别有效,因为这些微小的细胞已经通过辛勤劳作收获了湖水环境中的有机分子和营养物。这些细菌细胞因此能够给混合营养型原生生物提供高浓度高能量的套餐,正如它们给无色的鞭毛虫和纤毛虫等非光合原生生物提供的那样。

重要的循环

湖泊的微生物群系在食物网中担任多种角色,从生产者、寄生者到消费者,并且融入到动物的食物网中(图16)。此外,这些多样的微生物群落还在湖泊生态系统中的元素循环中占着举足轻重的地位,这是通过它们的氧化(失去电子)和还原(得到电子)反应实现的。这些生物地化过程不仅有学术上的意义,还完全改变了元素的营养价值、流动性甚至毒性。例如硫酸根是天然水体中氧化程度最高、硫元素最丰富的存在形式,也是浮游植物和水生植物为满足自身生化需求而吸收的离子。这些光合有机体将硫酸根还原为有机硫化合物。而在它们死亡并分解后,细菌会将这些化合物转化为臭鸡蛋味的气体硫化氢,这种气体对大多数水生生物都有毒。在无氧的水体和沉积物中,这种转化效应被硫还原细菌进一步放大,直接将硫酸根还原为硫化氢。还好另外一些细菌,即硫氧化菌,可以把硫化氢作为化学能能源,在充氧的水中将这种还原态的硫转化回到无害的、氧化态最高的硫酸盐形式。

碳循环是生态系统功能的核心(图17)。微生物在其中承担着完成许多关键转化的功能,但是矿物化学的作用也很重要。无机碳通过三种方式进入湖泊:气态的二氧化碳,碳酸氢根离子(HCO_3^-)和碳酸根离子(CO_3^{2-})。碳酸根离子主要来自流域中石灰石（碳酸钙）和白云石（碳酸镁和碳酸钙）的风化；二氧化碳从大气、入流和呼吸作用进入湖泊，后者往往来自细菌分解有机物质的过程。这三种形式之间有着如下化学平衡：

$$2H^+ + CO_3^{2-} \rightleftharpoons H^+ + HCO_3^- \rightleftharpoons H_2CO_3 \rightleftharpoons CO_2 + H_2O$$

这个平衡式对湖泊的酸碱平衡非常重要，因为碳酸根和碳酸氢根离子能够接受酸质子（氢离子H^+），这意味着进入湖泊的酸性物会很快被中和。但是，这种酸中和能力（或称"碱性"）在不同的湖泊之间差别很大。许多欧洲、北美洲和亚洲的湖泊因为没有足够的碳酸根来中和由大气工业污染引起的持续的酸雨，其pH值已经下降到了一个危险的低位。酸性环境对水生生物有负面作用，包括将铝元素转化为溶解度更高但毒性更强的离子形式，即Al^{3+}。幸运的是，在多数发达国家，这些工业排放已

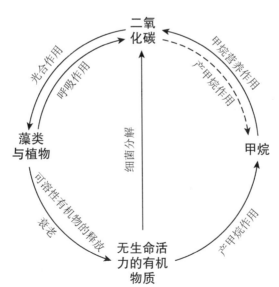

图17 水中的碳循环

经得到管制并减少。但是在某些流域,酸雨仍有持续影响,导致碳酸根离子和相应的钙离子长期缺乏。

光合微生物、浮游植物和水生植物会持续地消耗二氧化碳。为了弥补短缺和保持平衡,无机碳的化学反应平衡会被推向右边以补充被消耗的二氧化碳,而在这一过程中会消耗质子,从而引起pH值的上升。在高pH值的水中,尤其在水温较暖时,碳酸盐不再可溶并析出白垩色的悬浮物。这种现象被称为"白垩化"。国际空间站的宇航员们就拍到了北美五大湖颇为壮观的白垩化现象。

甲烷(CH_4)是水生碳循环中第二种极为重要的气体(图17),尤其是因为它像二氧化碳一样也是温室气体,而且每个分子的升温效果是后者的20多倍。甲烷的产生(即产甲烷作用)主要发生在无氧环境中,由一系列特定的古生菌引起,即湖泊微生物群系中的第三类细胞生命。这当中的微生物或用二氧化碳(图17中的虚线),或用有机小分子合成高还原性的甲烷。这些反应大多发生在有机物丰富的湖底的黑色软泥沉积物中,或者也可能在缺氧的底层湖水中,或者同时发生在湖泊的湖水和沉积物中。这些湖泊在冬季处于无氧状态。

湖泊在冬季产生气体的一些令人印象深刻的例子可以在北极苔原找到,这里分布着大量的热喀斯特湖。这些生态系统储藏着丰富的土壤有机碳,而这些有机碳是在冻土解冻或被腐蚀时进入湖泊的。它们在湖面冻结后迅速被细菌分解,同时造成氧气急剧下降,而冰面下的无氧湖水成为古生菌生产甲烷的理想生化反应器。在湖冰上钻个洞就可以看见气泡冒出来,这是冬季累积下来的甲烷。从这个口喷出的气体可以被点燃,在冰雪和湖面上产生壮观的火焰。

　　有机分子中的碳处于还原态，而碳价态最低的有机分子是甲烷。甲烷被氧化回到二氧化碳则会完成这一碳循环（图17）。大部分的颗粒或溶解有机碳都来自死去的藻类和植物，它们的分解过程由许多种可以分解各式各样的有机基质的细菌完成，同时这些细菌会从中获取大量能量。而甲烷的氧化则由少数的几种被称为"甲烷营养菌"的细菌完成。这些细菌一般只在氧气和甲烷混合的狭小区域出现。而值得一提的例外是夏季的热喀斯特湖。当覆冰融化时，大气中的氧气重新进入水体，这个同时充满甲烷和氧气的环境成为甲烷营养菌的天堂。一如所料，这些微生物组成了这些热喀斯特湖在夏季解冻时期细菌群的主体，有时超过总数的10%。

　　氮循环比碳循环复杂得多。这体现在循环涉及的分子和离子种类、氧化态以及参与循环的特定微生物上（图18）。氮气是大气中含量最多的气体，在湖泊中也一样。但是氮气分子（N_2）中的氮氮三键极度稳定，很难破坏。有一些蓝细菌可以通过酶催化完成这一过程，其中最出名的是引发水华的浮游植物长孢藻（之前被称为鱼腥藻），和能够分泌果冻一般的薄膜与球体的底栖属念珠藻。但总体来说，固氮作用并不是湖泊生态系统主要的氮源。例如，威斯康星州的门多塔湖是世界上被研究得最多的湖泊之一，最早由湖沼学家中的先驱爱德华·A.伯奇和昌西·朱代着手研究。这里每年都会有蓝细菌水华暴发，包括那些能够固氮的物种，但是通过生物固氮从大气中所得的氮还不足所有进入湖泊的氮的10%。

　　大部分进入门多塔湖的氮，同一般的湖泊那样，是从流域注

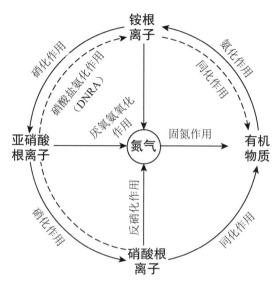

图18 水中的氮循环

入的入流、空气区（局部或区域性的大气）的降雨降雪以及风中携带的尘埃中获得的。这些氮是以硝酸根离子（NO_3^-）、铵根离子（NH_4^+）和其他各种各样的溶解或颗粒有机物的形式进入的。这些含氮物质有的在浮游植物细胞生长过程中被摄取吸收（即图18中的同化过程），而当细胞死亡，这些有机物质最终会被分解成有机氮和有机铵（氨化作用）。

有几个作用会将有机氮转换回氮气。铵根离子是浮游植物偏好摄食的氮形式，除此之外也能被一些特定的细菌（和一些古核生物）氧化成亚硝酸根离子（NO_2^-），随后被其他细菌氧化成硝酸根离子。有些硝化者叫"全程硝化菌"，能将铵根离子一路氧化到硝酸根离子。从这些离子的化学式就可以看出，每个氧化到最高价态的氮原子都需要三个氧原子，导致这些硝化者对氧气的需求极高。在富营养湖的底层水中，由于这种极高需求，不稳定的氧气平衡会被打破至彻底无氧状态。

在无氧条件下,其他细菌会主导氮的转换。有些菌种能够将硝酸根离子转为铵根离子,这一过程被称为硝酸盐氨化作用。它还有个更拗口的名字,异化性硝酸盐还原为铵(DNRA,见图18)。另一类对湖泊重要的细菌能够将硝酸根离子转为氮气,后者随之逸出并进入大气。这种反硝化作用会降低湖泊生态系统的氮总量。氮循环中还有一种特定的细菌叫作厌氧氨氧化细菌。这些把硝化作用和反硝化作用结合起来的细菌(浮霉菌门的成员)可以应用在水处理技术上,将含氮废水转为氮气排放到大气中。而且不像反硝化细菌,它们并不需要多少有机碳作为营养补给。

磷的生物地化循环包含另一套氧化和还原过程,对湖泊生态系统也很重要。该元素往往是藻类生长的限制因素,也是湖泊因生活垃圾和农业肥料发生富营养化的重要成因。不像碳和氮,磷没有气态形式,是以悬浮颗粒和溶解物的形式(如磷酸根离子和溶解有机磷)从流域的岩石和土壤中进入湖泊的。总磷量(TP)指所有溶解和颗粒态磷的总和,是衡量磷的指标,其浓度可从清澈的贫营养湖的 10 ppb(十亿分之一,又称十亿分率),到富含藻类的富营养湖的 100 ppb。

湖底的沉积物存有数量巨大的磷。尽管当中大部分都被隔离在表层湖水以外,有些在合适的条件下还是会释放出来并被运走。这种现象首次由杰出的湖沼学家克利福德·H. 莫蒂默发现并描述。他用英格兰湖区的温德米尔湖的沉积物做了一个经典的实验,将沉积物铺在一个水族箱的底部,并在上面灌满湖水,然后测量和氧气损失相关的化学变化。一旦氧气耗尽,沉积物中的溶解铁和磷酸盐就会大量进入水中。现在通过微电极可以更精确地展示这种现象。通过检测磷的新型电极可知，从北美五大湖中最浅的伊利湖采集来的沉积物，在有氧转为无氧的情况下，会使表层和上层水体的磷浓度上升一万倍（图19）。

沉积物表面的这些化学变化有几种机理。莫蒂默在猜想中正确地指出，含氧的沉积物中，大部分磷附着（吸附）在不溶性铁或氢氧化铁（Fe III）上。在无氧的情况下这些氢氧化铁被还原成可溶的亚铁价态（Fe II），这让磷得以释放到上层水体中。此外，在无氧状态下，还原硫酸根离子的细菌会产生硫离子，后者和铁结合，导致氢氧化铁吸收磷的能力下降。除此之外，含氧沉积物表面可能附着一层能够在内部以多磷酸盐颗粒隔离磷的细菌，而在无氧的条件下它们会将这些磷释放出来。当然其他的因素也会强烈影响这个生物地化反应，包括铝、钙、有机碳和

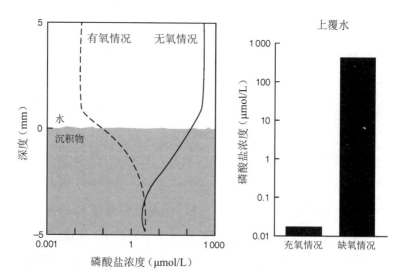

图19 伊利湖沉积物在充氧和缺氧（无氧）情况下磷的释放

pH。不出意料的是，这些因素对这一化学变化的影响在不同的湖之间相差巨大，有的甚至对无氧条件毫无反应。然而在一些富营养湖中，一旦这种有氧-无氧的临界值被突破（图19），从流域中进入湖泊后在沉积物中积累了几十年的磷就不再安定。这些被释放出来的磷会加速富营养化并阻碍改善水质的工作。

生产区域

由于当时的技术限制，福雷尔不可能意识到支撑湖泊自然循环的丰富微生物的多样性。而我们也是到了现在才了解到这些物种基因的丰富性，以及它们的关系所构成的复杂网络。但福雷尔意识到，根据食物网底端的光合群落，湖泊可以明确地分为两个区域。在近岸区域（即沿岸区），大部分的初级生产者都是水生植物，其中一些植根于沉积物中并完全被水淹没，而另一些则将叶子浮在水面或伸出水外。在较浅的湖泊和池塘，这些水生植物或称"水生大型植物"，以及它们附属的微生物会主导这一生态系统的总体生物生产。福雷尔对日内瓦湖繁茂的水生植物印象深刻，他在1904年写道："这些水生植物组成了真正的水下森林，就如同山岳间最美丽的森林一样美丽如画，神秘迷人。"这些"森林"是水生动物的重要栖息地，同时能够捕获营养物和影响水流。

在离岸更远的地方，即湖沼区或浮游区，它的美丽更多体现在微观层面：水中的浮游生物网包含着各种带有色素的细胞，它们有不同的尺寸、形状和颜色，其多样性令人震撼。这些藻类浮游生物或浮游植物能够捕获光以进行光合作用，一直到真光带的底部都能生长。这一界限也划分出底部以及沿岸区的离岸程度。尽管如此，对于单个植物种类来说，也会有其他因素限制其生长范围，如水压、基质种类以及食植动物如淡水龙虾。湖泊更深的地方属于深水区，在这里生命住在永久的黑暗之中，并依靠上层沉降的有机物质存活，尤其是那些从真光带不停沉降下来的浮游植物细胞（详见第五章）。

绝大多数湖泊都含有上百种浮游植物，主要可以归为四大类。首先是最重要的硅藻，它们拥有高度装饰性的硅玻璃外壳。这些藻类细胞可以通过标准显微镜观察和鉴别。福雷尔于1904年就它们在日内瓦湖食物网中的重要地位写道：

> 硅藻会被轮虫吃掉，轮虫随后被桡足虫吃掉，桡足虫又被水蚤吃掉，水蚤又被淡水白鱼吃掉，淡水白鱼又被狗鱼吃掉，狗鱼最后可能被人类或者水獭吃掉。

玻璃是一种重物质，因此许多硅藻受重力影响会下沉离开真光带，然后累积在沉积物中。硅藻的全年最佳生长时期在春秋两季。在这个时候，湖泊的充分混合运动能够使这些重型细胞保持悬浮在水中，同时阳光和营养物也很适合它们进行光合生产。世界上许多湖泊，包括温德米尔湖，都有长期的硅藻记录，这些记录显示了硅藻每年规律性的兴衰。这些硅藻季节性的死亡通常都是由于湖水开始分层并停止混合。浮游动物的捕食以及寄生生物的攻击则会进一步加快硅藻在下沉过程中的快速消亡。

在浮游植物中还有常见的非游动性绿藻。这些绿藻在大小和形状上相差巨大。最小的在2或3微米以下，被称为"微微型真核生物"。它们数量丰富，但通常需要DNA技术来鉴别。在另一个极端，有一些绿藻能够形成大的细胞集落。比如，只存在于日本琵琶湖的琵琶盘星藻三角变种，能够形成直径接近0.1毫米的集落。世界上许多湖泊的常见种球囊藻，也有类似甚至更大尺寸的凝胶状集落。这些集落颗粒如此巨大以至于浮游动物无法捕食，因此集落尺寸大能够有效抵御捕食者。

浮游植物的第三大类是可以游动的光合物种：植物鞭毛虫。这当中包括几个门的藻类以及许多物种，它们无论在生态学还是颜色上都不同。这些运动细胞通过鞭毛驱动自身游过黏稠的液态环境。其中一些物种如金棕色的分歧锥囊藻，有大小鞭毛各一个，能够形成跳动的树形细胞集落（图20）。其他的一些物种则有两个等大的鞭毛，如绿藻门的单衣藻。在这些能运动的

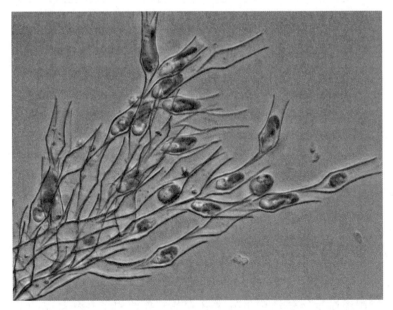

图20　植物鞭毛虫分歧锥囊藻集落。每个细胞介于10～15微米长，系混合营养型微生物：其既可以细菌为食，亦可利用阳光进行光合作用

浮游植物中的庞然大物当属棕色的双鞭毛虫门物种——飞燕角甲藻，广泛分布于世界上的许多湖泊中。它们的细胞长度可达250微米，能够每天在上层湖水中上下游动。

第四大类是蓝细菌。更广为人知的是它们之前的名字——"蓝绿藻"，得名于它们细胞内绿色的叶绿素和蓝色色素蛋白质结合呈现出的独特颜色。这当中包括超微型浮游种（微蓝细菌）以及能形成大型集落的物种如铜绿微囊藻。蓝细菌总体来说偏爱温暖的环境，所以它们的大型集落在晚夏到秋季尤其丰富，可能会导致水华暴发并严重影响水质。

浮游植物的总量和组成提供了有关湖泊生物生产力和水质的重要信息。最严谨的方法是在一个玻璃底的圆柱里灌满湖水并静置，然后用一个倒置显微镜透过玻璃观察沉积在玻璃片上的浮游植物。这种分析方法不但非常耗时，而且也要求显微镜学家有着高超的观察能力，以区分碎屑和藻类细胞，并鉴别藻类的种类。

另一种辅助性的方法是测量叶绿素a的含量。叶绿素a在所有浮游植物中都有，包括蓝细菌。我们可以通过高压液相色谱测量藻类的辅助色素，作为进一步估计浮游植物丰富度和其中包含哪些种类的依据。对大多数的样品来说，高压液相色谱会显示捕光色素的存在，如硅藻中的岩藻黄素和双鞭毛虫中的多甲藻素，以及各种各样保护细胞免受强光伤害的色素，如绿藻中的叶黄素和蓝细菌中的海胆酮。

这把我们带回到了哈钦森的悖论：这么多不同的种类是如何在浮游生物的微观世界中共存的？他给出了几种可能的解释，其中一个是这一群落可能处于不稳定的状态。由于湖泊的环境一直在变,今天的优胜物种到了明天就不再风光,这导致了混合的物种来不及在势头扭转之前把失败者完全排斥干净。这种观点在基因组分析的时代得到了进一步的发展。这种分析方法揭示了湖泊微生物群系中浮游生物的多样性远在哈钦森预料之外。微生物学家提出"稀有生物圈"的概念:包括浮游植物在内的大部分微生物物种的数量都少,在大部分时间内都生长缓慢甚至停滞,同时数量最多的物种也承受了最大的被病毒攻击和被摄食的压力。以一个湖泊的规模,即使每毫升湖水内只剩下几个细胞,整个湖泊中的细胞数量也还是巨大的,它们对冲了灭绝,也成为下次优势条件到来时的接种物。

第五章

终端是鱼的食物链

> 弱小者会成为强大者的猎物，后者又会被更强大的吞噬；或者如果它们逃过一劫，也无法避免微生物的降解。这是所有生命体直接或间接服从的规律。
>
> F. A. 福雷尔

弗朗索瓦·福雷尔在他的自传中回顾了他在湖沼学研究中最让他激动的时刻，也就是发现日内瓦湖湖底动物的时候。在位于莫尔日的家里分析离岸湖底沉积物中的波纹时，他首次察觉到它们的存在。当时他把一个沉积物样品放在显微镜下分析其成分，突然一个像蠕虫一样的生物闯入了他的眼中，并把眼前的矿物颗粒粗暴地推开。他对这种如此有生命力的东西感到震惊，然后马上开始猜测这种动物是不是栖息在日内瓦湖沉积物的最深处。假如是的话，那么"深水区并不是荒漠，而是一个深渊中的社会"。

那天晚上，福雷尔制作了一件底栖生物采集器，以获取日内瓦湖更深处的沉积物,而第二天的研究为他的学术生涯开辟了一片新的领域:他发现在深水区(图21)居住着各种各样的无脊椎动物,直至湖底300米左右深。这些居住在湖底的动物,或称"底栖"动物社群,依赖沉降的有机物质生存,尤其是从表层水沉至湖底的浮游植物。福雷尔把这种来自上层水的持续供给称为"他人餐桌的残羹冷炙",并且意识到底栖社群"收集了沉到底部的所有东西"。它们继而成为其他动物的食物供给,如在底部猎食的鱼类,同时沉积物中的细菌会将来自各处的有机物回收,分解为可溶解的营养物。大约在40年后,美国生态学家雷蒙德·L. 林德曼提及福雷尔在食物网、细菌分解和回收方面的"才华横溢的阐述"。在关于明尼苏达州赛达伯格湖的博士研究中,林德曼进一步建立并发展了这些理念,同时基于能量流和碳流定量地提出了"营养动力学"的概念,其中底层沉积物中的细菌和碎屑组成了连接食物网所有成分的枢纽。

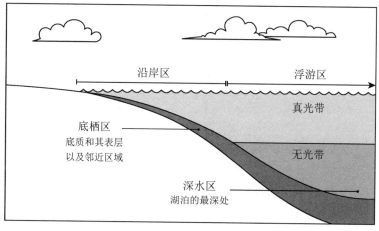

图21 湖泊的生态区

在水柱的上方,第四章所描述的生命支持系统(原生生物和细菌)不但为底栖社群提供碳和能量,还为浮游区或湖沼区的动物居民,即浮游动物提供能量。这些动物中体积最大的可以在湖水样品中观察到,是一些介于0.2～2毫米的微小游动个体,常常不连续地跳跃移动,而不是以一个方向在水中滑动。这些浮游动物会成为鱼类的食物,后者又会被其他生物吃掉,包括更大的鱼类、鸟类或人类。从高频声波和卫星遥感到脂质、同位素和基因分析,各种各样的新型技术与常规的观察方法相结合,让我们对浮游区和底栖区的食物网两者本身及两者之间的关系有了更好的理解。

水生食物网的最新研究成果让人们开始重视从周围流域进入湖泊的物质。这些物质来自陆生植物和土壤,为水生动物提供碳和能量的补充,并满足了它们在这方面的需求(图16)。外部环境对湖泊食物网的另外一种影响来自流域以外,即入侵物种。这当中的一些动植物是人为地被引入以"改善"生态系统,其他的进入则是意外,而且这种意外因人类的活动如划船和垂钓等而越来越频繁。许多情况下,这些入侵物种会剧烈地干扰原有的食物网,给湖泊提供的生态系统服务造成严重的破坏。

底部的生命

福雷尔把那个在他的显微样本中愤怒地蠕动的"可怜虫"鉴定为线虫或者圆虫的一种。实际上,湖泊的沉积物中栖息着三组重要的类蠕虫动物,归属于动物界中完全不同的门类。线虫是最丰富可能也是物种最多样的一类。这些像线一样的无脊椎动物通常长0.2～2毫米,在湖床中每平方米最多可分布100万只。这些最小的个体其数量也是最丰富的,大多数生活在沉积物上方几毫米间。人们已经发现近2 000个淡水线虫物种,但这个门类还没有得到很好的研究,因此人们估计还有数千种线虫亟待发现。它们在食性上也不尽相同,有些吃水生植物、原生生物和小型无脊椎动物,有些则是大型动物的寄生虫。另一些在湖泊沉积物中数量众多的线虫则以有机颗粒(碎屑)、细菌和微型真菌为食。

第二类是寡毛虫或称环节蠕虫。有些寡毛虫在湖泊的污泥和裂缝中移动与钻探,一些特定的种类如夹杂颤蚓(*Peloscolex variegatum*)只能在氧气含量很高的沉积物中存活。寡毛虫的一个主要亚类红线虫,能够分泌管状的黏液和颗粒,并垂直嵌在沉积物中。这种动物活在这些管道中,并把头埋进沉积物中进食,同时把尾部穿过管道伸入上层水中摆动以吸氧。这些寡毛虫因为带血红色的色素所以颜色鲜艳,能帮助它们在低氧的环境中存活。最典型的两种是正颤蚓(*Tubifex tubifex*)和霍甫水丝蚓(*Limnodrilus hoffmeisteri*),常在受有机物污染的沉积物中出现,是水质不良的指标。

湖泊沉积物中类蠕虫动物的第三大类实际上是昆虫的幼虫形态,特别是蝇类昆虫(双翅目)。最常见的幼虫是不咬人的蠓虫摇蚊科,包括5 000多种已知的种类。这些幼虫是底栖鱼类和其他动物喜爱的食物,而且分布密度极高,在湖泊沉积物中每平方米可达上万只。由于体积比线虫大很多,摇蚊常常在底栖群落中就生物质而言占主导地位。它们当中许多物种都有喂食管,且能用身体运动制造水流把充氧水拉过来。这种挖掘活动让它们成为"生态工程师",能够极大地改变沉积物中的氧气条件和生物地球化学性质。如同其他大多数底栖动物,这类生物的物种多样性和数量在沿岸区最多（图21），这归功于这里的基质多样性、植物和藻类的碎屑以及从流域进入的有机物质。但是对中型到大型湖泊来说，沿岸区的面积相对于深水区很小，所以深水区的湖泊总生物质也许会更大。

摇蚊在湖泊科学中占有很重要的地位，这是因为德国普伦水生生物实验室的主任，杰出的动物学家奥古斯特·蒂内曼对这种动物的研究最为感兴趣。他的一位国际同事，水生植物学家埃纳尔·瑙曼，在瑞典隆德建立了湖沼学研究所的科研站，并基于藻类浓度发展了一套湖泊的分类方法。他把这称为湖泊的营养状态（trophic state），源于希腊语 *trophikos*，意为"营养物"。他将湖水分为贫营养（水质清澈，浮游生物数量少）和富营养（浮游生物数量多）两大类。蒂内曼采纳了瑙曼这一在今日广泛使用的分类方法，并提出在这两种营养状态下摇蚊群落的组成完全不同。例如长跗摇蚊属在贫营养湖中很常见，而摇蚊属会在低氧的富营养湖中出现。两位湖泊科学家于1922年在德国基尔举行的一个成立大会上联手创立了国际湖沼学学会（SIL）。

软体动物是另一组在湖底数量丰富的动物，有两个分支：腹足纲（即腹足类）与双壳纲（即蛤蜊和贻贝）。一些特定的鱼类会吃腹足类动物，后者会在沿岸区的水生植物间寻找避难所。在这个栖息地，它们会吃碎屑和覆在植物及沿岸区湖底的藻类生物膜（即附着生物）。蛤蜊以它们的生命周期长而闻名，可达几十年，而部分珠蚌的分布机制令人印象深刻并以此闻名。这些蛤蜊有可以伸出壳外的鳃膜，外形如小鱼一般（有时甚至含有色素，可假装成眼睛）。这片鳃膜会随蚌壳一起跳动并引诱捕食的鱼类前来。一旦鱼类张口咬下，鳃膜会突然张开并释出叫"钩介幼虫"

的繁殖体。这些幼虫会寄生在毫无警觉的鱼的鳃上，然后长成小蚌，直至足够重并最终从鱼鳃跌入远离其亲本所在的沉积物中。

　　第三种构成大部分底栖动物生物质的是端足类动物，也被称为淡水虾或"飞毛腿"。它们是甲壳亚门的清道夫，大多以碎屑为食，有近2 000种在淡水中生活。虽然看上去只有一个物种在主导着群落，DNA分析却表明，它们当中有许多隐藏的种类，外形相同但基因不同。在贝加尔湖，人们可以观察到最大规模的适应性辐射。这里至今鉴别出了260种地方特有物种，另有80个亚种，估计还有数百种亟待鉴别。福雷尔发现，在日内瓦湖深水区的沉积物上有一种"盲虾"很常见，其拉丁文名为 *Niphargus forelii*。这一物种之后在这个湖中灭绝了，但仍能在瑞士、德国和意大利的其他高山深水湖中找到。在北美五大湖中，糠虾可以占据底栖无脊椎动物总生物质的50%。而在日本琵琶湖，地方特有物种安氏汲钩虾的种群密度可达每平方米6.3万只。

　　除了蠕虫、软体动物和端足类动物外，还有许多其他动物生活在湖泊的底栖栖息地，但通常所占生物质较低。这其中包括

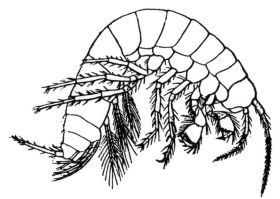

图22　日内瓦湖的盲虾（*Niphargus forelii*）

小型物种如轮虫和水螨，除端足类以外的甲壳亚门如介形纲（沙虾），桡足亚纲中的猛水蚤目以及特定的一些枝角目，尤其是盘肠溞科。螯虾产于各种淡水底栖栖息地并有当地的叫法。比如新西兰的毛利人把湖泊中的淡水螯虾称为"koura"，而澳大利亚的原住民将滑螯虾称为"yabby"，在美国南部则通称螯虾，并被当地人养殖，是"卡真"菜系中的重要食材。螯虾是杂食性的，大部分都生活在沿岸区，并以植物、腹足类、摇蚊、蜉蝣和碎屑为食，然后它们又会被鱼类和鸟类吃掉。淡水海绵存在于许多湖泊中，底栖区还生活着许多淡水水母的水螅体，附着在水下植物和其他基质上。正如福雷尔惊奇地发现的那样，即便是最深湖泊的湖底也肯定"不是荒漠"，而仍是生物丰富和具有动物生产力的地方。

浮游生物的关系网

在湖泊的敞水区，有三类浮游动物在把食物网底层（如浮游植物与细菌）的碳和能量传递到上层鱼类的过程中发挥了重要作用，它们是轮虫、枝角类动物和桡足类动物。这些类别中的第一个在动物界中独占一门，即轮虫门。这是根据它们的线形纤毛所组成的、如车轮一般旋转的双冠状外观（纤毛冠）命名的。这些纤毛在水中驱动轮虫并将食物颗粒引导至嘴中。这些动物首次由显微镜学的先驱安东尼·范·列文虎克在一滴池塘水中发现，他将它们命名为"轮形微动物"。轮虫在海里很罕见，但在淡水中单就数量来说是最丰富的浮游动物，如在北部苔原的热喀斯特湖，它们的种群密度可达每升1 500只。轮虫一般很小（小于0.2毫米），生活周期很短，通常只有几天。大部分轮虫以微型藻类、其他原生生物和细菌为食，但也有肉食种，如晶囊轮属。这些动物转而又会被桡足类动物和幼鱼吃掉。

第二组浮游动物是枝角类动物,是介于0.5～2毫米之间的甲壳亚门动物。它们当中包括86个属,其中只有4个属生活在海里。最常见的3种浮游属是溞属(也称水蚤,但它们完全不像跳蚤一样是寄生虫)、象鼻溞属和单枝溞属,后者有着标志性的巨大果冻状外壳,以保护头部及抵御捕食者的攻击。一般来说,这组浮游动物是小型"食浮"鱼类最喜欢的食物。被称为盘肠溞科的枝角类动物广泛分布于沿岸区,和水生植物及沉积物有着密切联系。枝角类动物的身体包裹着由甲壳素组成的外骨骼,随着它们成长而脱壳掉落,这一过程在某些物种中可以进行超过20次。在一些非常清澈的湖泊中,它们的外壳中可能含有黑色素(如图23所示),作为防晒剂保护这些动物和它们的卵免

图23 芬兰一湖泊中的浮游动物影溞(*Daphnia umbra*)的显微成像。每个个体约2毫米长

于紫外线辐射的破坏。

枝角类动物有多对附肢,每一对都有专门的功能。最重要的附肢是用于游泳的触角。水对这种体积的动物来说是黏稠的介质,而这些触角能起到桨一样的作用。它们的腿(4～6对)上有细微的毛发(刚毛),毛发上又有更细的毛,叫作"纤毛"。纤毛能够通过某种方式如静电作用过滤水中的颗粒。这些食物包括藻类、细菌、其他原生生物和碎屑。收集来的食物材料会被另一对触角尝试,被其他附肢(颚)碾碎,并与黏液一起被搓成球(食团)再送入口中或丢弃。

和轮虫一样,枝角类动物能在很短的时间内实现惊人的种群增长,这是"孤雌生殖"的结果(图24)。通常在临时的池塘和水池中会有凭空出现的涌群。这些种群大部分的个体都是雌性,可以无性产卵(无须受精),并在育卵室中孕育胚胎,最终释出能自由游动的"新生儿"。根据物种和食物条件的不同,单个母体可能携带从1到超过200个卵,而在暖水中胚胎的发育时间

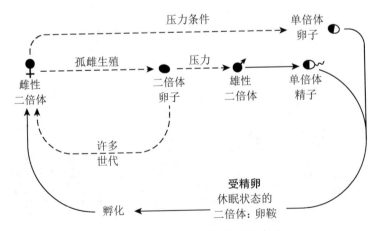

图24 枝角类浮游动物的无性(孤雌生殖)和有性生殖

可能只要两天。

这种无性繁殖策略对在稳定环境中的快速生长非常有效。但和一些轮虫一样，枝角类动物在条件恶化的时候更偏向于有性生殖（图24）。这可能是由物理压力引起的，如极端温度，也可能由生物压力引起，如拥挤和食物短缺。这时雌性会产下只含一套染色体组的单倍体卵和会孵化成雄性的双倍体卵（两套染色体组）。这些雄性会和雌性配对并让它们的单倍体卵受精，成为双倍体合子。许多物种都会将这些合子携带在改造过的外壳中，然后在蜕壳的过程中将它们释放出来。这些合子是黑色裹起来的休眠卵，被称为"卵鞍"。卵鞍对极端环境有很高的忍耐度，如干燥和冰冻环境，而且可能对它们在水体间通过风或鸟羽传播有着重要意义。它们能保持数月、数十年乃至上百年的休眠状态，例如在一个干涸的池塘底部，一旦有利条件恢复，便可孵化成无性繁殖的双倍体雌性。

桡足类动物是海洋浮游动物中数量最丰富的一类，在几乎所有湖泊中都很常见。在大型深水湖泊如贝加尔湖和北美五大湖中，就生物质而言它们是浮游动物中最多的。和枝角类动物一样，桡足类动物也属于甲壳亚门，有由甲壳素组成的外骨骼和多对用于游泳、觅食和感应的附肢（图25）。和枝角类动物不同的是，它们没有无性繁殖的阶段，种群由雄性和雌性混合组成。在交配过后，雌性产下会孵化成幼虫或"无节幼体"的卵。这些卵要经历5～6次的蜕壳才可以成为桡足类动物的幼虫（即"桡足幼体"）。随后它们会经历5次蜕壳并成为性成熟的成年体。在暖水中，它们的生命周期可能在一周内完成，但在极地或高山湖泊的冷水中可能需要至少一年。桡足类动物以浮游植物和其

图25 加拿大魁北克市南部一湖泊中的桡足类浮游动物 *Aglaodiaptomus leptopus* 的显微成像。这种动物长2.3毫米

他原生生物为食,是食浮鱼类富含脂质的食物来源。但是相比行动缓慢的枝角类动物,桡足类动物更难被捕获,因此在食浮鱼类众多的湖泊中它们占有统治地位。

游来游去

轮虫、枝角类动物和桡足类动物都是浮游生物,因此它们的分布受到水流和湖泊混合作用的强烈影响。但是它们都能游动,并且可以调节自己在水中的深度。对于最小的浮游动物来说,如轮虫和桡足类动物,这种游动能力有限;但大一点的浮游动物能够在24小时的昼夜循环中游动相当长的距离。福雷尔在夜间划船到日内瓦湖中采集浮游生物样本时首次观察到这种现象。他发现他的拖网会捕获"浮到水面大量的切甲亚纲动物［桡足类动物］"。之后的研究显示，日内瓦湖的枝角类动物白天活动在温跃层和深水层，入夜后则向上游约10米；而桡足类动物则会向上游60米，在日间又返回深邃黑暗寒冷的深水区。

　　更大的动物会在水柱中移动更远的距离。负鼠虾是一种糠虾，可长达25毫米。它们于日间居住在湖底。而在太浩湖的夜晚，除非有月光，它们会上游数百米至湖面。贝加尔湖中主要的浮游动物之一是一种地方特有的端足类动物，拉丁文学名为 *Macrohectopus branickii*，可以长至38毫米长。它们能在白天距湖面100～200米的深处形成密集的群落，但在夜间会分散并浮至水面。这些夜间迁徙活动把湖泊生态系统中深水区和浮游区的水面连接起来。人们认为这是浮游动物为了在日间规避依靠视觉捕猎的捕食者（尤其是浮游区的鱼类），同时在夜色掩护下获得水面的食物而演化出的习性。

　　鱼类对迁徙规律最显著的影响之一体现在一种被称为"幽灵蚊"的昆虫上。这种昆虫有长达2厘米的类蚊幼虫，也被称为"玻璃虫"，因为它们的身体是透明的，并在两端有一对帮助它们浮在水面的空气囊。在欧洲，这种昆虫主要包含两个物种，各自有不同的昼夜移动规律。黄库蚊（*Chaoborus flavicans*）主要生存在有鱼的湖泊和池塘中，日间待在水底，以沉积物中的动物为食。它们能进行基于苹果酸的无氧代谢活动，能够在无氧的水中或者将头埋在无氧的沉积物中存活。在夜间它们会迁徙到湖面捕食浮游动物，尤其是桡足类动物。这种迁徙规律受鱼类出没的影响最为明显，它们似乎能够通过化学信号（有鱼腥味的利它素）侦测鱼类的存在。第二种是暗库蚊（*Chaoborus obsuripes*），会避开有鱼类的水体，并在整个昼夜循环中待在靠近水面的区域，即底栖捕食者（如蜻蜓幼虫）的活动区域以外。

尽管某些鱼类会在湖泊特定的区域活动，也有其他鱼类能够在区域之间游动并活跃在不同的栖息地。例如，在世界上面积最大的淡水湖苏必利尔湖（82 100平方千米，最大深度406米）中，湖白鲑大部分时候都是食浮性的，出没在近岸敞水区。但是在晚秋会迁徙到湖边产卵，其富含脂质的鱼卵为沿岸区的生态系统提供了能量补贴，为鲱形白鲑（湖白鱼）贡献了34%的能量需求，后者大多时候以底栖猎物为食，如近岸浅水中的端足类动物。这种鱼类迁徙意味着湖泊生态系统的不同部分在生态上连接了起来。

对于许多鱼类来说，其穿梭的栖息地范围可一路延伸到海洋。溯河洄游的鱼类[①]每年会迁出湖泊并游入海洋。尽管这种迁移会造成大量的能量消耗，它的优势却在于能为鱼类带来丰富的海洋食物来源，同时为鱼苗的成长减少被捕食的压力。北极红点鲑即是一个例子，它们生活在大不列颠寒冷的深水湖以及包括日内瓦湖在内的欧洲湖泊中。这是一种生活范围最靠北部的淡水鱼类，一直到加拿大高北极地区北纬83°的湖泊A中都能发现它们的身影。它们生活在淡水中的时候主要以底栖无脊椎动物、浮游动物、小型鱼类和水面昆虫为食，而在海里则吃其

---

① 指在海洋中生长，成熟后上溯至江河中上游繁殖的鱼类。——编注

他鱼类和端足类动物。现在，结合声呐标记法和基因标记法，人们可以标记捕获的鱼类并将它们释放，从而鉴别这种鱼类的不同种群，以及迁徙的来源地。

降河洄游的鱼类有着相反的迁徙规律，会在淡水中生活而在海水中产卵。其中一个例子是欧洲鳗鲡，它在河流中最为丰富，但在自然和人工湖泊中也很常见。卫星标记法显示成年欧洲鳗鲡会迁徙5 000千米甚至更远去到马尾藻海产卵。它们以每天10～30千米的速度游动，这是一个漫长的过程，可能长达一年，并会因为被猎食而在数量上遭到严重的损失。鱼苗之后会通过墨西哥湾暖流和北大西洋暖流返回至欧洲的水域。

所食即所是？

饮食习惯对湖泊食物网中动物的营养状态绝对有着重要的影响。但就具体的物种而言，其特殊的需求和生理状态也有变化。以简单的元素比值为例。著名的美国海洋学家阿尔弗雷德·C. 雷德菲尔德提出了以下理论：海洋中的颗粒（大部分由浮游植物组成）的碳氮磷原子量比一般来说为106∶6∶1（或41∶7∶1的质量比）。他也提到在深海中再生的营养物有着同样的氮磷比。湖泊中的浮游植物和其他颗粒的碳氮磷比接近或稍高于雷德菲尔德的比值，但在食物网的上层，不同类型的动物的碳氮磷比也很不一样。桡足类动物的氮磷质量比一般在14∶1之上，但即使是同一个湖内的枝角类动物也往往只有这个值的一半，约为7∶1。枝角类动物这种显著的低比例是因为它们细胞内合成蛋白质的细胞器数量更多，尤其是含有RNA的核糖体——一种富含磷的生物分子。高磷成分会促进生长，但也导致对磷的生物需求更高,也就使得氮磷比更低。对湖泊营养成分比例的分析被称为"生态化学计量学",为湖泊科学带来了重要的见解和问题,包括不同动物群体对磷的不同需求以及这种不同对营养物循环的影响。

食物数量和供应速率对食物网中的所有动物都很重要。通常来说,湖泊中动物生产力("次级生产力")随着浮游植物及其生物质的光合生产("初级生产力")上升而上升。但这不仅仅关乎数量,也关乎质量。这些食物的养分之间有着巨大的不同。如食物中油脂(脂质)的组成对动物的健康、繁殖以及生存有着重大影响。这些脂质分子有的有着明亮的颜色,这是藻类色素经动物摄食转化而来的(如图25中桡足类动物的亮橙色是由类胡萝卜素的一种——虾青素引起的)。对于某些浮游动物,如清澈的高山湖泊中的种类,这种色素可能主要用于抵御紫外线辐射,但对其他种类来说这似乎是维持高能量脂肪贮存的一种方式,以用于度过冬天并在下一个春天生长和繁殖。

有些脂质分子被称为"多不饱和脂肪酸"(PUFA),如 $\omega-3$ 多不饱和脂肪酸和二十碳五烯酸(EPA),对性激素的分泌尤为重要,后者能调节包括人体在内的身体功能,如脑部发育、视觉和心血管代谢。大多数水生生物都不能分泌PUFA,而必须从饮食中摄取,最终从生产这些PUFA的藻类和摄食它们的消费者中摄取。湖泊科学家对在食物网中追踪EPA和相关PUFA有着浓厚的兴趣,因为这反映了水生生态系统中的摄食关系以及系统的健康状态。有的PUFA甚至能够通过水生昆虫(如摇蚊和蜻蜓)被鸟类捕食从而转移至陆生物种(其PUFA含量总体少于水生生物)上。脂质对理解化学污染的影响也有关系。多数有机污染物（如农药）都是脂溶性的，因此可以随着脂质在食物网中转移，并在食物链的顶端（或更高营养级）动物处浓缩（即生物放大作用）。

分析食物网最有力的方法是基于自然存在同位素的分析。同位素是两类拥有相同质子数的原子，因此属于同种元素，但有不同中子数。比如空气中的氮气，大多由两个有七个质子和七个中子的氮原子组成，被称为氮-14（化学符号为 ^{14}N），但有一小部分（0.366 3%）含有多一个中子的氮原子，被称为氮-15（化学符号为 ^{15}N）。氮被动物吸收后，^{15}N 会比 ^{14}N 在动物体内保留得多一点，这种富集效应在食物链中逐步向上延续。

单个中子产生的差异似乎微小，但借助灵敏的质谱仪，即使是微小的 ^{15}N 富集也能被准确地检测出来。例如，在贝加尔湖的浮游区，每在食物网中向上一步，^{15}N 就会富集 3.3 ppt（图 26）。贝加尔湖的大型地方性硅藻——贝加尔湖直链硅藻——能够摄取无机氮。它们的 $\delta^{15}N$，即硅藻中的 $^{15}N/^{14}N$ 比和大气中这一比值的差值，约为 4 ppt。硅藻会被端足类动物吃掉，并将氮在食物链中一路向上传送，最终到海豹时 $\delta^{15}N$ 为 14 ppt。其他氮来源会带来变数，但这种方法对研究食物网中"谁吃谁"的关系提供了宝贵的参考。对肉食性动物来说，这种方法还可以研究不同食物在其饮食中的比重。碳的自然同位素之比（$^{13}C/^{12}C$）也可以用于测量饮食来源。而在海洋研究中，硫的同位素之比（$^{34}S/^{32}S$）也能起到相似的示踪效果。在水蒸发从液相转变为气相的过程中也会发生同位素分馏现象，所以氢（$^{2}H/^{1}H$）和氧（$^{18}O/^{16}O$）在水分子中的同位素比在水文学中用于测定湖的蒸发降水平衡。

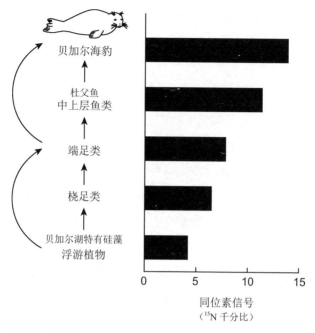

图26 贝加尔湖的中上层水体食物网和对应的氮-15信号（$\delta^{15}N$）在各营养级的递增

湖中的入侵者

在19世纪末，福雷尔警觉地观察到来自加拿大的水蕴藻进入了日内瓦湖。这种水草在整个湖泊中"茂盛而可怕地扩张着"。这一入侵物种是人为引入的，以改善本地池塘和溪流中的鱼类栖息地。但就如同世界其他地方不幸地发生过的那样，这种水草很快就进入湖泊的沿岸区并在其中扩张。这个物种和水鳖科的其他水生植物于20世纪中叶侵入新西兰的湖泊，在水下形成高达6米的森林，极大地改变了沿岸栖息地，并影响了水库电站的发电量。其他物种如穗状狐尾藻，一种欧亚水生耆草，给饮用水源（包括魁北克的圣查尔斯湖）造成了一些问题。原产南美的水生风信子——凤眼莲，是一种浮在水面的侵略性植物，会将水面覆盖致使水生栖息地窒息，在亚洲、非洲和美国南部的湖泊肆虐，包括维多利亚湖的近岸区域。

一种入侵动物的到来对湖泊的巨大影响首先体现在食物网的一层，然后扩散到整个食物网中。这种"营养级联"的经典案例发生在弗拉特黑德湖，一个在美国蒙大拿州的大型深水湖（占地500平方千米，最大水深116米）。1968—1975年间，一种叫 Mysis diluviana 的糠虾（和欧洲的孤糠虾是近亲）被引入弗拉特黑德湖上游的三个湖以改善鲑鱼渔业。到了1981年，这种虾顺流而下进入了弗拉特黑德湖，并于1980年代末在数量上经历了爆炸性的增长（图27）。在糠虾进入弗拉特黑德湖的几年后，湖中浮游动物中的枝角类动物和桡足类动物因众多糠虾的过量捕

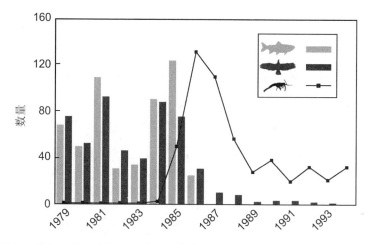

图27 糠虾入侵后弗拉特黑德湖的食物网变化。淡水红鲑（图中数值乘以100）和秃头鹰（图中数值乘以7）的数量系鲑鱼上游产卵地测得，而糠虾数量（图中数值乘以1 000）系湖中每平方米水柱所含数量

食而消失殆尽。随之而来的是食物网自上而下的效应：因为被浮游动物摄食的压力减小，浮游植物的生物质大量增加，其群落组成也发生变化。

淡水红鲑也是弗拉特黑德湖的引进种，它们变得没有浮游动物可吃。又因为糠虾只有在夜间才出没于浮游区的水面，红鲑无法看见它们因而不能摄食。这个出乎意料的规避鱼的习性让引入糠虾成为提升鲑鱼产量的一个糟糕的选择。在弗拉特黑德湖的流域，鲑鱼的竞技性捕捞量从1985年的超过10万条直线下滑到1988年的0条。秃头鹰会在淡水红鲑产卵的溪流聚集捕食红鲑，其数量从1980年代早期的600余只，到十年后的基本绝迹（图27）。另一个食物网效应是糠虾成了另一个引入种湖鳟（湖红点鲑）的主要食物来源，因为后者在湖底觅食。湖鳟的崛起正在把原生种公牛鳟（强壮红点鲑）逼向灭绝。

一旦进入湖泊就能造成最大破坏，也是最成功的入侵物种都有以下几个特点：生长速度快，忍耐宽泛，在高种群密度下能茁壮成长，以及有通过人类活动增强的迁徙和繁殖能力。斑马贻贝在这几项上得分都很高，在世界各地都是引起诸多麻烦的入侵物种。它们原产于里海地区，但随着18和19世纪全欧洲兴建运河，它们很快传播开来，并在1824年抵达大不列颠。1988年它们首次出现在北美五大湖，人们认为它们如许多其他入侵物种那样，是通过货船的压舱水入侵的。到1990年它们遍布五大湖，现在迁入密西西比河的河床。它们的近亲斑驴贻贝也在大约相同的时间入侵了五大湖，在松软的沉积物和比斑马贻贝所处更深的水域中繁殖，造成了其他的问题。

一个斑马贻贝可以在产卵季节产下一百万颗卵，随后它们会孵化成可以自由游动达一个月的分散幼虫（面盘幼体）。成年贻贝的密度可以达到每平方米上万只。它们在水管中的繁殖造成了核电站和热电站冷却系统的严重问题，同样给饮用水厂的进水口带来了麻烦。一只斑马贻贝一天可以过滤一升水，把其中的细菌和原生生物吃得一干二净。在伊利湖，斑马贻贝入侵后不久，湖水澄清度变为原来的两倍，硅藻减少了80%～90%，伴随着浮游动物的下降以及对食浮鱼类的潜在影响。这个物种的入侵能将一个湖泊的食物网从由浮游区主导变为底栖区主导，但代价是本地的蚌科蛤蜊会在斑马贻贝群中被闷死。它们高效的过滤能力也会导致初级生产者水文情势的转变，从高浮游植物浓度的浑浊湖水转变为底栖水生植物主导的澄清的湖泊生态系统。

  现在，入侵物种带来的问题在全球气候变化的背景下变得更为复杂了。这会削弱本地植物、动物和微生物在它们温度范围上限的竞争力，并为之前在温带和热带区域生活的物种开辟了新的栖息地，使它们可以扩张到之前寒冷且不宜生存的湖泊。保护区（如国家和地区公园）对湖泊食物网的保护有着前所未有的意义，能帮助它们抵御额外的压力，并减少入侵物种带来的压力，这种压力不可避免地伴随着土地开发和相关运输路线扩张而来。

第六章

极端湖泊

> 我们所认识的表面湖水的组成,到底是在所有深度上都一致,还是会变化?以什么规律变化?
>
> F. A. 福雷尔

我们对湖泊的上层水域很熟悉,但这对我们认识下层水域往往没有什么帮助。在某些湖泊中,这些不同深度的水域有着极端的差异。南极洲麦克默多干谷地区的万达湖是最明显的一个例子。这里的湖面终年结冰,因此阻止了风致混合。当第一批科学家在冰面上钻洞,并将热敏电阻探头伸入下方水柱时,他们惊奇地发现温度在随着深度上升,并在底部达到26℃。这种暖水在冷水之下的反向温度分布是由明显的盐浓度梯度分布引起的:万达湖的表层湖水是纯净的冰川淡水,而底层水的盐度是海水的三倍。经过一段时间的激烈辩论后,这种出乎意料的温暖被解释为阳光的累积作用。夏季的阳光辐射透过剔透的冰和澄清的淡水,逐渐加热底层致密的、高盐度的水。年复一年,一个世纪又一个世纪,这种累积最终达到今天观察到的不同寻常的温度。

极端湖泊是指有着不寻常的物理、化学和生物特质的水体,它们具有极大的科学价值。世界上许多地方都分布着咸水湖,且通常生产力极高,只靠一条简化的食物链就可以支撑一大群的候鸟和留鸟。极地和高山湖泊受冰雪的强烈影响,因此对水冻融阈值内温度的微小变化很敏感。这些高纬度和高海拔的生态系统是过去和现在全球气候变化的前哨,也是深入理解湖泊微生物学和生物地质化学的模型。世界上其他极端湖泊还包括酸性湖、碱性湖、地热湖以及那些周期性喷发的湖泊,它们会释放出液态或气态物质,危及附近的人类。

在大多数条件极端的湖泊中,只有最顽强的、酷爱极端环境的微生物才能在其中生存和成长。这些"嗜极微生物"包括酷爱极高盐度的嗜盐微生物、适应了终年寒冷水体的嗜寒微生物(psychrophiles,来自希腊语 *psukhrós*,意为"寒冷"或"冻结"),以及在低 pH 下生长最好的嗜酸微生物。对这些微生物的生物化学和基因研究为地球生命的起源、进化和极限提供了深刻的见解,并催生了独特的生物分子的医药和生物技术应用。

咸水湖

水的众多显著性质之一是其介电常数异常地高,这意味着水是一种强极性溶剂,有着能够稳定进入溶液离子的正负电荷。这种介电性质是由分子中不对称的电子云造成的,如在第三章中所描述的那样。这让水在经过地面的时候能够将土壤和岩石中的矿物质过滤出来,并把这些盐保持在溶剂中,甚至在浓度很高的时候也是如此。

所有这些溶解矿物质共同产生了水的盐度。盐度是指每升水中溶解的盐或固体的克数。因为每升水重1千克,盐度也可以表示成"克每千克"或"ppt"。海水约为35 ppt, 其盐度主要来自阳离子如钠离子(Na^+)、钾离子(K^+)、镁离子(Mg^{2+})和钙离子(Ca^{2+}),以及负离子如氯离子(Cl^-)、硫酸根离子(SO_4^{2-})和碳酸根离子(CO_3^{2-})。

这些溶质统称为"主要离子",能够导电,因此一种测量盐度的简单方法就是测量浸在溶液中的两个有着固定间距的电极间的电导。现在湖泊和海洋科学家的常规测量活动包括使用温盐深测量仪测量描绘盐度-温度关系图。这种水下仪器能够拴在一根绳子上在水柱中下降,并在每秒内多次记录电导、温度和深度。电导的单位是西门子(或微西门子,记作μS,鉴于淡水湖中盐浓度比较低),会校准到标准温度25℃并给出特定的电导率,单位为μS/cm(微西门子每厘米)。

所有淡水湖都含有溶解矿物质,具体的电导率介于50～500 μS/cm之间;而咸水湖的值能超过海水(约50 000 μS/cm),这些水体是极端微生物的栖息地,如嗜盐绿藻杜氏藻和耐盐古核生物(嗜盐古菌),它们拥有的生化策略能够应对如此高盐度带来的压力。南极洲西福尔丘的迪普湖盐度非常高(270 ppt),以至于湖水在隆冬季节也不结冰,可以在周围天寒地冻的湖心划船。但最好远离湖水,因为这些液态盐卤的温度约为-18℃。

世界上咸水湖总面积巨大,并保持了几个世界湖泊记录。世界上最大的湖是里海,面积超过37.1万平方千米,最大深度1 025米。其盐度属于中等偏高(12 ppt),大部分来自陆地而不是海洋。与之相反的是黑海，因为和地中海交换水所以被认为是海洋系统而不是湖泊。和许多咸水湖一样，里海是一个古老的水体，因占据地质活动所形成的构造盆地而成。那里有许多地方特有的物种，包括一种内陆海豹——里海海豹。世界上最古老的湖泊是盐度中等（6 ppt）的咸水湖伊塞克湖（意为"温暖的湖"），坐落在吉尔吉斯斯坦的天山地区。这个大型深水湖（占地6 300平方千米，最大深度702米）有可以匹敌贝加尔湖的湖龄（约为2 500万年），为多种动物群提供了栖息地，包括地方特有物种。在海平面400米以下的死海是世界上海拔最低的湖，也是最咸的湖之一。其盐度约为342 ppt，约是海水的10倍。在极端高海拔地区也有咸水湖，包括青藏高原以及玻利维亚和秘鲁交界的高原。虽然有着各种不同寻常的特点，但因为它们常常地处偏远以及水质咸涩不可饮用，咸水湖一直被人们视为可有可无。但是，它们对候鸟的价值和稀有生物的重要性，使得它们在世界上好几个地区都处于保护区争夺的前线。

在决定是否拯救加利福尼亚州莫诺湖的过程中产生的争论，是关于咸水湖价值争论最为典型的案例。这一争论历程漫长但最终取得胜利。当马克·吐温在1860年代早期游览这个地区时，他把这个湖称为一块"可憎沙漠"中一片"庄严、宁静、无人航行的海"。但如同许多咸水水体一样，莫诺湖是一处令人惊叹的美丽之地，栖息着大量的浮游生物和水鸟。湖水被认为是"三系水"，即其盐度主要来自三种成分：碳酸盐（因此也被认为是碱湖）、氯盐和硫酸盐。假如你把手伸入湖水，会发现湖水有种滑滑的肥皂水的感觉。当湖水在沙漠的阳光下迅速挥发，你的手上会留下一层盐膜，就像一只薄薄的白手套。

当地下泉水流进莫诺湖,这些含有钙离子的冷淡水遇到咸湖水时,碳酸钙以石灰石的形式沉积出来,并形成被称作钙华塔的石柱。附在钙华塔上的蓝细菌也会促进这一过程,光合作用消耗的二氧化碳会将平衡推向碳酸盐析出的一端。许多令人印象深刻的钙华塔随着湖面在古今的下降而暴露出来,有些高达数米(图28)。

莫诺湖坐落在加利福尼亚州内华达山脉的东侧,位于广阔荒凉、高高隆起、干旱贫瘠的大盆地边缘。这个地区曾经存在面积广阔的淡水湖,但在古老的湖水蒸发后就剩下盐碱地和咸水湖。这些残余水体中最大的是犹他州的大盐湖(4 400平方千米,最大深度14米),盐度在50 ppt ~ 270 ppt之间,取决于波动的水位。大盐湖、莫诺湖、里海和其他许多咸水湖都是"内陆湖",意味着它们没有出流。莫诺湖的巨大蒸发损失被每年来自

图28 加利福尼亚州莫诺湖的钙华塔

内华达山脉融雪所补充的淡水抵消。但洛杉矶的水利规划者为了满足城市人口快速增长的需求付出了巨大的努力,他们将内华达山脉的融雪分流,通过引水渠将水引至560千米外的城市。第一次主要的分流从1941年开始,也是自那时起莫诺湖的规模开始缩小,其盐度也在上升。从1940年代到1970年代,湖的水位下降了约15米,盐度从40 ppt翻倍到80 ppt。洛杉矶市的用水已经导致区域内另一内陆湖欧文斯湖的彻底干涸,而莫诺湖似乎也走在相似的消失道路上。

1976年,一群来自加州大学戴维斯分校和斯坦福大学的本科生在莫诺湖边参加一个科研夏令营,以研究湖泊的生态,莫诺湖的命运也自此逆转过来。他们的研究对象是湖泊中的食物网,其生产力高度发达主要归功于两种顽强的无脊椎动物:一种是在湖边的水域度过其幼虫期和蛹期的碱蝇,它们是库泽狄卡原住民过去的食材;另一种是每年在湖中数目能达到兆级的丰年虾(卤虫),主要以一种微型(小于3微米)耐盐的绿藻(*Picocystis*)为食。

莫诺湖的学生研究显示,这些碱蝇和丰年虾是每年夏天在这个湖中转的大量候鸟的食物来源。这其中包括5万只海鸥、8万只瓣蹼鹬、超过100万只黑颈䴙䴘和许多其他种类。最重要的是,他们发现盐度的上升会导致丰年虾的灭绝。食物骤减,再加上水位的下降会引发湖心岛和湖岸的相连,让筑巢的鸟类(如加利福尼亚海鸥)暴露在北美野狼和其他捕食者的面前。学生团体成员在鸟类学家戴维·A. 盖因斯的领导下成立了莫诺湖委员会,将洛杉矶市告上了法庭,引起了公众对莫诺湖生态系统的生态价值和悲惨命运的关注。经过在法庭上长达15年的马拉松式斗争和不懈努力，委员会终于胜诉并取得了法律和政治上的支持，湖水入流得到全面恢复，湖面水位得以上升。莫诺湖现在是一个独一无二的保护公园，每年吸引着许多游客和大量的候鸟前来。

极地和高山湖泊

高纬度和高海拔的湖泊包括一系列各种各样的生态系统，从北极河流三角洲洪泛平原上兴衰的湖群，到加拿大北部广袤深邃的大熊湖（面积31 153平方千米，最大深度446米），再到有着高度分层水体的南极洲万达湖，以及清澈深邃的高山湖如比利牛斯山脉中的勒东湖（海拔2 240米，最大深度73米）。勒东湖最初由杰出的加泰罗尼亚生态学家拉蒙·马加勒夫所研究。虽然是不同种类的栖息地，极地和高山湖泊还是有几个共同的特点，包括远离城市的地理位置和污染物对其直接的影响。这些特点让这种湖泊成为追踪重金属和有机污染物远程传播的理想场所。例如在勒东湖，有机污染物如滴滴涕和多氯化联苯在欧洲禁用几十年之后还能在湖水中检测出来。这说明这些污染物是从数千千米以外的地方过来的，而控制它们对生物圈的毒害需要全球的通力合作。其他的污染物被检测出是从特定区域过来的，如六氯环己烷在南欧用作农业上的杀虫剂。极地和高山湖泊地处偏远的特点引起了生物地质学的研究兴趣。一方面，有证据显示一些耐寒微生物在全世界都有分布，如某些淡水蓝细菌；另一方面，其他研究显示这些如岛屿一般孤立的生态系统有着区域性的微生物集合，包括蓝细菌和单细胞真核生物（原生生物）。

极地和高山湖泊的另一个共同特点是它们和冰雪圈的密切联系,后者即世界上所有含雪含冰的环境的集合。这些湖泊在一年大部分甚至所有时间内都覆有厚冰。冰盖上常有积雪,会限制可供初级生产的光线。在极地,这种效应又进一步被每年历时三个月的长夜所加剧。气候变暖对这些湖泊的一个主要影响是无冰期的延长,即在春天融冰的时间更早,在夏季结冰的时间更晚。这不仅为光合作用提供了更多的光线,也将供给藻类生长的营养物向上混合至表层湖水。但是,在水面敞开的情况下,湖泊生物群也会更多地暴露在潜在有害的紫外线辐射之下,可能会影响生产力和种群组成。

在极地和高山湖泊中存活非常成功的一组生物是"耐寒"蓝细菌。它们能够忍受极端寒冷甚至完全冻结的环境,但在更高温度下生长也更快(因此不像"喜寒"生物那样对寒冷有嗜好)。这些微生物把沙子和碎屑颗粒固定在它们的丝状体和由糖类化合物组成的黏液中,形成了厚实的生物膜,或称"微生物垫",覆盖在湖泊、池塘和溪流的底部。这些微生物垫常常呈亮粉或亮橙色,这是由其中的胡萝卜素引起的,以保护其免于明亮阳光中紫外线B的伤害。它们的厚度介于零点几毫米到数十厘米之间。最壮观的种群可以在永久覆冰的南极洲湖泊的底部找到,如温特塞湖。它们形成的拱形结构外观很像地球最早的化石(叠层石)。

在寒冷覆冰的水中,蓝细菌菌垫和菌膜经常与苔藓共存,有着高浓度的红色和蓝色蛋白质。这些蛋白质被称为"藻胆蛋白",能够高效地捕捉光以进行光合作用。这些种群在极地和高山湖泊的初级生产和生物质中占主导位置。对这些微生物集群构成的分析显示,尽管蓝细菌占主要地位,但还有其他成千上万的微生物存在,如古核生物、病毒和真核生物(如硅藻和小型无脊椎动物等)。这些真核生物以此为栖息地,享用着蓝细菌菌垫中丰富的营养物。在7.2亿～6.35亿年前的雪球冰河期,冰川覆盖了几乎整个地球。冰面融化形成的池塘底部覆有生物膜,有证据显示,这成为真核细胞(原生生物)的避难所(如同今天它们在极地冰盖和冰川上所做的那样)。

极地和高山湖泊也是研究及更好理解湖泊元素循环运行机制的有效系统模型,并且可以用于研究湖泊是如何被周围环境的物质输入所影响的。对这些研究最有帮助的自然实验室是那些在极地地区永久分层的湖泊,如南极洲麦克默多干谷的万达湖、弗里克塞尔湖、霍尔湖、邦尼湖、乔伊斯湖和米尔斯湖。这些水体被称作"局部循环湖",意思是它们只是部分混合。另一个分布着这些永久分层的咸水湖的区域是加拿大的高纬度极地地区。这些湖泊首先在1969年由一支军事科研探险队发现,并以战术命名的方式把它们称作湖泊A、湖泊B、湖泊C等等。这些湖泊仍保有着这些带有冷战时期色彩的无聊名字,但字母掩盖了它们许多不同寻常和有趣的特点。

湖泊A(北纬83°,最大深度128米)坐落在叫作*Quttinirpaaq*的国家公园里,这在因纽特语中意为"世界之顶"。这里有一座峡谷,在5 000年前曾是连接北冰洋的一个充满海水的峡湾。随着北极冰盖的融化和相应上层巨大压力的下降,这一峡谷从海中升起,将峡湾孤立成含有北冰洋海水的潟湖。融化的雪和冰川成为淡水,流入湖中并浮在致密的咸水上。湖泊A的分层现象现在可以通过其盐度分布(图29)观察到:低电导率的融水出

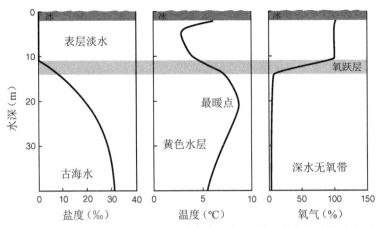

图29 加拿大高纬度极地湖泊A的湖水在盐度、温度和氧气含量上的分层

现在冰面下的表层,而在11米处盐度急剧上升,并随水深增加保持这一趋势,直至于几千年前就困在峡谷的古海水。

湖泊A的盐度分布为其地质历史提供了线索,而温度分布则记录着最近的变化。图29中展示的温度分布显示了冰面下的湖水在夏季变暖的现象,这很可能与温暖低盐的入流相关。之后温度随水深下降,但在约22米深的地方意外升高,并达到最高值。和南极洲的万达湖一样,这种温度的上升是穿透冰层和水层到达这一深度的阳光对湖水逐渐加热的结果。

湖泊A表层湖水的溶解氧是饱和的,在夏季时由从融化雪堆流出的溪流补充。但在更深的水域,含氧量在氧跃层突然下降直至低于检测限值,这种无氧的环境一直延伸到湖底。湖水颜色和气味随深度的急剧变化是分层现象的进一步证据。例如在28～30米深处有一个黄色水带,这是绿色光合硫细菌造成的,它能够捕捉光能并用硫化氢还原二氧化碳成糖(而不是像植物那样用水还原),并在此过程中产生黄色的硫单质颗粒。在30

米及以下采集上来的水样品都有一股硫化氢的臭鸡蛋味。

基于核酸（DNA 和 RNA）的分子工具为分析不同水层的微生物群落提供了强大的方法，现在已经常规地用于湖泊研究中。这些工具为研究微生物多样性和生物地化过程提供了深刻的见解。因为水生微生物群落的绝大多数成员都不能被培养，也不能在显微镜下进行区分，所以这些特点在之前都不可研究。一旦 DNA 被分离出来，且其中的核苷酸被测序（核酸的 A、G、C、T 等字母顺序），通过绘制生命树图即可探索它们的遗传关系，其中的样品或物种间的距离即是关系远近的衡量指标。这种方法的其中一个优点是所有数据都在国际数据库（基因银行）中共享。这一数据库可以提供巨大的（目前已有两亿条记录）且永远在增长的数据参考源以便测序对比。

这种分子方法应用在湖泊 A 上的一个例子如图 30 所示。首先要从 10～12 米深处采集水样品，这个深度是氧含量急剧下降的地方（图 29）。此处往往是寻找新的微生物的地方，因为在氧浓度梯度上一般含有各种各样的氧化剂和还原剂，满足了微生物的各种生活方式。人们发现核糖体基因的 DNA 测序对鉴别不同物种特别有用，因此对它们进行了比较。湖泊 A 的 DNA 结果显示，三种湖泊微生物都聚集在生物树上的古核生物分支中（图 30 中的 0.05 比例尺代表 5% 的基因差异），而且它们和海洋氨氧化古菌亲缘关系较近，后者是一种能够将铵根离子氧化成硝酸根离子的古核生物。在湖泊 A 深 10～12 米处的化学环境对基于氨氧化反应的能量生产很有利，因为氧气可以从氧跃层上方扩散而来，同时铵根离子则从下方的无氧区扩散而来。

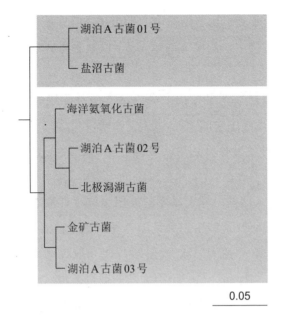

图30　湖泊A中三种古核生物的遗传亲缘关系树状图，以及它们与其他栖息地古核生物的亲缘关系

基于DNA的方法对冰下湖的研究很有价值，这种湖泊是所有极地水生栖息地中条件最严苛的。湖泊的水体常年保持液态，但深埋在南极洲冰盖几百甚至几千米以下。第一个冰下湖在苏联位于南磁极附近建立的沃斯托克科研站下方被毫无征兆地发现了，当时恰逢国际地球物理年（1957—1958）。无线电回声探测显示，在这个科研站所在的3 750米厚的冰盖下有一个深达1 000米的液态水层。后续的地球物理学测量揭示，这个被称为沃斯托克湖的隐藏水体有着巨大的面积，占地14 000平方千米，预测体积有5 400立方千米。这个体量远远超过世界上许多其他巨大的湖，如安大略湖（1 640立方千米）。这一被冰隔绝的水体的发现提出了一个科学界和公众都深感兴趣的问题：沃斯

托克湖是一个处于构造湖盆中的无菌水体吗？还是一个在缺乏了数千年支持生命的阳光的情况下，以某种方式活跃的湖泊生态系统？

在沃斯托克湖里寻找生命的这一任务激励着天体生物学家，他们对地球生命的起源、进化、极限以及那些可能允许地外生命存在的环境有着浓厚的兴趣。人们已经在太阳系中的其他地方发现了液态水，如木星最小的卫星木卫二，和土星第六大卫星土卫二，这些星球上厚厚的冰层下都存在液态水。沃斯托克湖似乎是研究这些生态系统的合适对象。同时，人们可以在此发展出一套无菌的钻冰采样技术，以获取这些环境中的化学样品甚至生物样品。另一个更强烈的动机来自对南极冰盖的探索。人们发现有数以百计的冰下湖（大部分都比沃斯托克湖小得多），其中多数都在有流动水联通的盆地中，其面积和亚马孙流域一般大，藏在厚厚的冰层下。这让冰下水体成为世界上最大的生态系统类型之一，可能对下游的沿海南大洋有着重要影响，这些水最终汇入其中。地球物理学家对这些冰下液态环境有着浓厚的兴趣，因为冰盖和陆地交界处起润滑作用的水会影响冰盖的稳定和流动性，这对全球气候、洋流和海平面有着重要的影响。

从冰下湖中采样的最初几次尝试被一些挫折和不确定性阻碍了。俄罗斯已经在沃斯托克站钻到一个相当大的深度，获得了过去气候变化的记录。他们那3.4千米长的冰芯给温室气体在过去40万年以来的自然循环提供了前所未有的视角，表明当今人类活动带来的温室气体已经大幅超过过去的最高值。但是，在长达10年的钻探过程中，人们用航空煤油保持钻洞敞开，最终在2012年2月成功破冰至水中时，所采得的样品很可能被这种液体污染了，导致很难分析本地微生物群落。2012年12月，一个英国科研团队尝试采样埃尔斯沃斯湖，一个150米深、覆冰3 400米、位于南极洲西部的冰下湖。他们用无菌的热水钻探系统以确保不会污染冰下水，但不幸的是在钻探300米后燃料就耗尽了。

2013年1月，一个美国团队利用热水钻探技术和一系列方案，钻探进位于南极洲西部的惠兰斯湖，同时避免了微生物和化学污染。根据覆冰的水平位置变化，人们知道这个湖有着规律的干涸和充水过程。在采样时，它有2.2米深，并覆有800米厚的冰。这个团队运用DNA测序方法得到微生物的群落结构，发现水中有各种各样的微生物，以氨氧化古核生物和一种只存在于北极永久冻土中的硝酸盐氧化细菌为主，前者同样存在于湖泊A的氧跃层中。他们也钻得一个沉积物芯，其中的微生物包括生活在表层的消耗甲烷的细菌，和在底层的生产甲烷的古核生物。

还有许多问题亟待解答，如冰下湖是否存在由真核细胞和病毒参与的生物互动所构成的微生物网络，又如惠兰斯湖微生物的代表性如何。但是，这些初步结果为证明冰下环境是一个巨大的活跃生态系统提供了有力的证据。这个生态系统包含那些以无机化学物为能量来源的微生物，以及其他从有机材料中获得能量的微生物。在未来的几十年里，南极冰下湖的探索将会继续成为"极端湖沼学"中一个激动人心的前沿，给生命如何在冰期时覆盖了地球大部分面积的巨大冰盖下存活这一问题提供了见解。

喷发性湖泊

作为极地与高山区域冷水生态系统的另一个极端,地热水也吸引着湖泊科学家和微生物学家的浓厚兴趣。在这里,生命又一次地被推向其生存的极限。但出人意料的是,有各种各样的嗜极微生物能在低pH和灼热高温这些恶劣环境下生存。上述的基因技术也成功应用在了这些水域,揭示了不同寻常的物种的存在,以及它们在这些严酷栖息地的生存策略。在这些湖泊生活的微生物中发现的一些生物分子被证明有巨大的商业价值。对这些地热生态系统的微生物进行生物勘探,实现了生物科技和生物医药产业中新产品的开发。最知名的产品是一种叫作"Taq聚合酶"的酶。"Taq"是指水生嗜热菌,一种首先从美国黄石国家公园的热水池中分离出来的微生物,也是Taq聚合酶的来源,用于在分析中放大DNA信号,即聚合酶链反应(PCR)。自然中的水生嗜热菌生活在50～80℃的水中,所以其热稳定的DNA聚合酶在PCR中的交替高温下也能理想运作。

在活跃地热区域生活要面对一个常年威胁,即地面及与其相连的水不时地有喷发的趋势。这可能是以水热爆炸坑的形式发生的,其中包括水蒸气在内的被困住气体的气压最终超过了它们周围岩石和土壤的压力阻力,遂从地表喷出。留下的大洞填满水,成为湖泊。活火山口也可能充满水形成湖泊,这些湖水可能在火山爆发时被喷出,或因溢出的火山灰而干涸。

这种湖的一个例子是图31中位于鲁阿佩胡山的火山湖。这个湖的湖水温度波动很大,有时可达60℃;酸度也很高,其pH值可低至0.9。在过去的150年中,这座火山经历了三次大型喷

图31 位于新西兰鲁阿佩胡活火山口的强酸性湖泊

发,伴随着越来越频繁的较小型的喷发。这座火山正受密切监视,在其白雪皑皑的山坡上装有地震预警系统,以警告滑雪者在有喷发的时候迅速躲到安全的位置,避开可能从山谷冲击下来的浑浊湖水和沉积物(火山泥流)。这个监测机制是发生在1953年12月23日的一起悲剧促成的。当时鲁阿佩胡火山湖在经历过一次爆发之后,湖水突破了拦灰坝的限制,火山泥流从一个河谷中流出,冲毁了铁道干道上的一架桥。由于对这个数分钟前发生的灾难毫无预警,一列夜间快车的火车头和前六节车厢掉进了山谷中,导致151位乘客死亡。

火山湖能对人类产生其他威胁,包括它们释放的过饱和气体所带来的危害。喀麦隆的尼奥斯湖盘踞在一座死火山的火山口,但在湖底下的一个岩浆腔将二氧化碳泄漏到了湖水中。这些浓度极高的二氧化碳会在滑坡或地震的过程中突然排放到大气中。1986年，一朵巨大的二氧化碳云从湖中逸出，导致周围地区1 746人和3 500头牲畜窒息而死。从那时开始，人们在湖中插入管道以将深水中的气体排出，从而降低突然爆发的风险。另外一个气体累积的相似案例发生在莫瑙恩湖，同样也是位于喀麦隆。1984年的一次爆发逸出了大量的二氧化碳，导致37人窒息而死。

　　第三个面积更大、充满火山气体的湖是基伍湖，位于卢旺达和刚果民主共和国边境。这个大型深水湖（2 700平方千米，最大深度480米）的底部湖水由于和一座火山相互作用，积累了大量的甲烷和二氧化碳。这些气体时不时地从湖中逃逸。这种有毒的富含二氧化碳的气体在斯瓦里语中被称为 *mazuku*，意为"邪恶的风"。福祸相依，深处的甲烷同样是可以用于发电的潜在燃料。现湖边已经装上一个装置，这个装置将水抽起并汲取其中的甲烷用于燃烧。这能产生2 600万瓦的电力，同时降低了水中的气体含量，以及在深水灾难性爆发的风险。

第七章

湖泊与我们

> 人类对自然和栖息者的影响比其他任何动物都要大。
>
> F. A. 福雷尔

当弗朗索瓦·福雷尔开始给日内瓦湖的动植物分类时,他列入名单的第一个物种就是智人。他提出,人类不仅通过自身活动,如湖岸开发、货运和客运(图32),而成为湖泊生态系统的一部分,也有能力对湖泊和其提供的服务,如渔业和安全饮用水,造成巨大的破坏。他观察到,和自然成因一样,人为的干预也能改变湖水水位。他同时也是状告日内瓦市的专家级证人,控诉其对日内瓦湖出流的管理不善。他几乎意识不到的是,在20世纪,拦坝造湖会成为人类社会的主流,并且当今在发展中国家继续受到狂热的拥护。

福雷尔调查了日内瓦湖广阔的水域(面积580平方千米,容积89立方千米),认为日内瓦湖会给湖泊居民提供无限量的优质饮用水。但是在20世纪晚期,如同世界上许多其他湖泊那样,

图32　19世纪日内瓦湖上的传统商船

日内瓦湖开始经历富营养化，伴随着水质的极速恶化、底部溶解氧的耗尽以及藻类的生长。对于日内瓦湖和所有淡水资源来说，最大的挑战可能还在前方，即全球气候变化及其相关现象，如温度升高、混合模式的变化、极端天气事件、供水的变化和本地及入侵物种栖息条件的改变等。

大大小小的水坝

几千年来，人类一直在截坝并把水围困成人工湖泊和池塘。直到19世纪晚期，所有这些拦坝成湖的行为都是小规模的，并带有各种用途的结构，包括灌溉农作物、喂养家畜、管控洪水、供给饮用水、满足文化和审美目的、为水车提供动力以及蓄养鱼类等。而在20世纪，为引航和水电而建造的大规模水坝成了进步的象征，在湖水水域扩张的同时带来了巨大的经济效益。欧洲的水库现今总共占地10万平方千米，包括伏尔加河上的两个坝前水库，古比雪夫水库（6 450平方千米）和雷宾斯克水库（4 450平方千米）。世界大坝注册中心现记录着58 519个"大型水库"，即其大坝高度在15米及以上的水库。这些水库总共贮存了16 120立方千米的水，等于在美加边境的尼加拉瓜大瀑布213年的流量。世界上最大的水电设施之一是加拿大魁北克北部的詹姆斯湾设施，它从1980年代晚期开始运行，有一个占地11 800平方千米的水库，能够产生16 500兆瓦的电量，并且现在仍在扩建。

尽管在西方世界，建坝的趋势已经放缓甚至逆转，但在亚洲、非洲和南美洲，这种活动正如火如荼。中国长江上的三峡大坝（水库占地1 084平方千米，坝高181米）从2012年开始运行，以发电量计（22 500兆瓦）是世界上最大的水电站。在非洲，大约有100个大型水坝正在规划或建造中，包括位于蓝色尼罗河的145米高的大埃塞俄比亚复兴坝。在南美洲的亚马孙盆地，有超过300座大坝正在规划或建造中，包括欣古河上的贝卢蒙蒂大坝设施。

水库有几个和自然湖泊不同的特征。首先，它们的湖盆形状（形态）很少是圆形或椭圆形的，往往是树杈形的，有着树状的主干和分支伸入淹没的河谷。其次，水库的流域与湖面面积比往往很高，这反映了其来源为河流。对自然湖泊来说，这一比值相对较低。如对英格兰湖区的温德米尔湖和沃斯特湖来说，这一比值是16左右；日内瓦湖则为13.8；而太浩湖只有2.6，这一因素导致太浩湖湖水有着漫长的滞留时间（650年）。与之截然相反的是，对于圣查尔斯湖而言，即魁北克截坝而成的饮用水水库，流域湖面面积比是46；美国卡罗拉多河上胡佛大坝前的米德湖是640；三峡大坝水库则是923。这些相对较大的流域意味着水库中水的保留时间很短，水质相比在没有激流的情况下要好得多。然而，在封闭的河湾、支流和靠近大坝的下游还是会发生有害的藻类水华。

相比自然湖泊，水库的水位波动往往更大也更快，这限制了沿岸区动植物的发展。水库另一个明显不同的特点是它们的各种条件沿河流呈梯度分布。在上游，河流部分的水是流动、湍急且充分混合的。然后水会经过一个过渡区，到达坝前的湖泊部分。这里往往是水库最深的地方，分层现象更明显，水质也更澄清，因为来自陆地的颗粒会下沉。在一些水库中，湖水的出流位于大坝的底部，即深水区。这减少了氧气耗尽和养分累积的程度，同时也为大坝下游的鱼类和其他动物群落提供了冷水。如今，人们对泄流的时机和规模越来越谨慎，以维持这些下游的生态系统。

为了满足水电、灌溉和饮用水的需要，大坝创造了新的湖泊，但是其环境代价往往在当时并不总是显而易见的。奥卢米耶湖是伊朗的一个大型咸水湖（最广阔时占地5 200平方千米），以其中生活的鸟类而闻名。由于在其三个主要入流处都因水电和灌溉的需求建了大坝，湖泊的面积缩小到原来的10%。这导致沉积的盐碱被风吹散，影响周围农田和人类的健康。广阔的咸海（乌兹别克斯坦和哈萨克斯坦交界）也遭遇了类似的环境问题。由于上流被引流用于灌溉，其面积由1960年代的68 000平方千米缩减到2005年的7 000平方千米。现在在其北部角落建起了一个水坝，以保留湖水、淡化盐碱以及恢复原有湖盆小部分的渔业。

建坝可能对河流盆地原有的居民造成广泛的影响，包括人类和动物。贝卢蒙蒂计划会淹没数以千计的亚马孙印第安人使用的土地,其文化影响已经引来国际社会的关注和抗议。为了建造三峡大坝,约有120万人被迁置,包括13个城镇的所有人口。现在大坝阻止了动物的迁徙,包括中华鲟和其他濒危鱼类。但是最大的影响可能还是在下游:船只运输量增加,同时每年的洪泛地区也被改变。有证据表明,长江洪泛平原上的水位降低会加剧寄生扁虫从水生软体动物到人类的传播,导致严重的"蜗牛热"或血吸虫病;而这一疾病也在埃及阿斯旺大坝建造后变得普遍。较低的水位会给湿地带来危险,也削减了鱼类和其他水生动物的栖息地之间的联系。洪泛区域的改变还可能会影响本地物种适应河水自然循环而进行的产卵、孵化、生长和迁徙活动。热带河流盆地的大坝对鱼类多样性的影响尤其引人关注,如亚马孙河、刚果河和湄公河,这些地方栖息着约4 200个物种,其中60%是本地物种。这三个河流盆地加起来共约有840个大坝正在运作或建造,另有445个尚在规划中。

大坝对下游的影响会一直延伸到海洋。由于沉积物和营养物被保留在水库中,海洋的食物网获得量就少了。这种减少也可能会改变海岸线,引起滨海三角洲的退化和海水的倒灌,因为自然的侵蚀过程已经无法被来自上游的沉积物的补充所抵消了。自三峡大坝运行以来,人们已经观察到长江三角洲的严重侵蚀。水库的另一个影响是被淹没的植物和土地中的汞会进入水体,然后被细菌甲基化成甲基汞。这种毒性更高的物质会在食物链上的每一环累积,最终被输送至海水。

世界上许多人都依赖水库以管控洪水、供应用水、生产电力和提升经济效益。这些截坝而成的湖所提供的生态系统服务现在也成为我们文明的一部分。世界上大部分地区都在继续建造大坝，同时也有人呼吁增加大坝的建造，以减轻气候变化对未来水资源可用程度的影响，减少对化石燃料的依赖，以及跟上到21世纪末还要增加30亿的全球人口不断增长的需求。然而，历史在友好地提醒我们，这些项目的代价往往会被低估，而效益会被高估，对人类和环境影响考虑不周，长此以往会损害社会和生态价值。

世界淡水变绿

全世界范围内的湖泊所面临最严重的问题是富营养化引起的藻类和水生植物过度繁殖，富营养化即由人类活动所导致的水体中营养物过多的现象。这一问题在20世纪中叶浮现在人们面前。当时人们意识到，湖泊会逐渐变得富营养化，透明度下降，最终被沉积物和生长的植物所填满。这种缓慢的自然过程被周围流域的人类活动所带来的养分输入大大加快了。其所导致的"富营养化"或"超富营养化"（即养分丰度更高）水域被媒体称为"死湖"。虽然这些湖中的有毒藻类和氧气缺乏会带来死亡和灭绝，但是这个词是误称。因为富营养化的湖水中还是有水生生物的，只不过这些生物主要由有害的物种组成，会严重损害渔业、饮水和其他生态系统服务。

营养物的富集可能由"点源"导致，即从管道排放到接受水体的污水；也可能由"非点源"导致，如来自道路、停车场、农用地、漫灌田的径流，以及开垦地上被清理掉的含水和营养物的植被。在1970年代，即便世界上最大的湖泊也开始显现令人担忧的先兆：日益增加的养分富集导致了水质恶化。如在日内瓦湖，福雷尔在1870年代测得的冬季西奇盘深度为15～20米，1970年代跌至最好时的10米。在福雷尔的报告中，哪怕是在分层现象往下300米处，日内瓦湖的底层湖水氧气含量还是很高的。但在100年后，深水的氧气浓度已经下降到一个缺氧的值，即2毫克每升以下，把底栖动物逐出湖底的一部分区域，并可能会导致部分物种如丰年虾的灭绝（图22）。

水质透明度的极速下降往往是富营养化的最初迹象，虽然在森林地区的湖泊，这种迹象可能会被湖水中溶解的有色有机物对光的强烈吸收掩盖很多年。在分层期，底部湖水氧气含量的突然下降是富营养化的另一个指标，最终会导致分层以下的湖水彻底不含氧气（无氧状态）。但是，富营养化对生态系统服务最显著的影响是有害藻类水华的暴发，特别是蓝细菌引起的水华。

在富营养化的温带湖泊中，通常有四个属的蓝细菌会形成水华：微囊藻属、长孢藻属（正式名字为鱼腥藻属）、束丝藻属和浮丝藻属。它们或单独或集体暴发。尽管有着独一无二的尺寸、形状和生活习性，它们还是有一些令人印象深刻的共同生物特征。首先最重要的是，它们的细胞内通常都充满了憎水的蛋白质荚子，会排出水并留住气体。这些充满气体的蜂窝状结构被称为"气囊"，会降低细胞的密度，允许它们浮至水面接受生长所需要的光线。

把产生水华的一滴湖水放在显微镜下就可以马上看到，每一个单独的细胞体积都非常小，而水华时每升湖水中含有数以十亿计的细胞。在图33所示的例子中，每个细胞直径约为5微米，并有明显的亮点，那代表着能够散射光线的气囊。对于如此

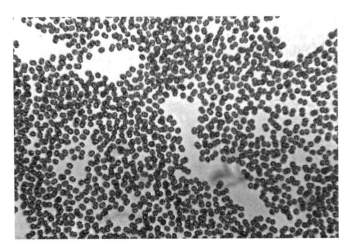

图33 有毒水华铜绿微囊藻的显微成像

微小、孤独的细胞,其上浮速度非常慢以至于可以忽略,但当它们形成多细胞的菌落之后,集体的浮力可以让它们的上升速度非常快,如微囊藻菌落每小时可以上升5米。

这种上浮能力也可以调节。在白天,这些细胞捕捉阳光并通过光合作用合成糖,这增加了它们的密度,直到比它们所处的水重,导致它们下沉到营养物更丰富的下层水柱或沉积物表层。这些糖会被细胞内的呼吸作用消耗掉。这种物质损失使得细胞密度下降,到比水低的时候它们就会再次浮起。这种下沉上浮的交替会导致表面水华在24小时昼夜循环内大幅波动。

形成水华的蓝细菌在湖面的累积会带来湖面污泥,这些污泥会被冲刷到湖湾和沙滩上。在水柱中,尤其是在上层水柱,那些种群致密的群落会把底栖植物所需要的光线遮蔽,并大幅减少其他浮游植物的光照。其所导致的"蓝细菌的绝对优势"和藻类物种多样性的下降对水生食物网有着负面影响,特别是因为浮游动物难以过滤和消化这些大型群落。除此之外，蓝细菌往往缺乏脂肪酸，所以对动物来说也是劣质食材。又因为水华消散被分解时需要大量消耗氧气，对食物网的负面影响可能会变得更为复杂。

蓝细菌水华给饮用水设施的管理带来了极大的麻烦。首先，由于生物质的过量生产，大量藻类颗粒会超出水处理设施中过滤系统的负载能力，尤其是当它的入水口被置于港湾中或者群落悬浮的某一深度时。其次，水的味道也会受影响。蓝细菌通常会合成各种各样的次级产物。次级产物是指生物的衍生化学物，它们不参加如光合作用、呼吸作用和生长等初级作用。对于许多甚至绝大多数次级产物，人们还不知道为什么蓝细菌会费力合成它们。尽管具体原因未知，猜测和假说却不少。这其中包括浮游植物种类之间进行的化学战争、通过毒性抵御食植动物的摄食、对痕量金属的调动以及进行细胞间的沟通等。这些生物化合物中有一些会产生令人不快的味道和臭味，包括带有土腥味的土臭味素和2-甲基异莰醇、带有草腥味的环柠檬醛以及在分解过程中产生的带硫臭味的烷基硫化物。最后也是蓝细菌最重要的影响，是这些次级产物有的毒性很高。

有毒的湖

2014年8月2日星期六，俄亥俄州托莱多市的市长D. 迈克尔·科林斯紧急召开新闻发布会，宣布居民不应饮用或煮沸自来水。同时，在有进一步通知前，全市所有餐厅都须关门。市水处理设施的环境化学家在他们常规的水质化验中检测到一种蓝细菌毒素含量突然升高，其名为微囊藻毒素-LR，超出了世界卫生组织限定的 1 ppb。托莱多市的水是从伊利湖中抽取的,在这里每年都会大面积地发生蓝细菌水华。水华中常含有这种毒素,有时其浓度会超过世卫组织规定值的 100 倍。但是,微囊藻毒素-LR 大部分只存在于细胞内,一般可以通过过滤掉藻类颗粒去除。托莱多市的问题在于这种毒素进入了处理后或说"润色过"的水中。在进行进一步的检测和安全程序后,市政府在随后的那个星期一解除了自来水使用限制。但是在一个有 50 万居住人口的城市关闭自来水,这对美国和加拿大造成了持久的公共影响,并将富营养化和有毒湖水的严重性重新摆在了公众面前。

蓝细菌毒素的问题在世界其他地方也引起了极大的关注。太湖是中国第三大湖,也是一千万人口的饮用水来源。尽管面积巨大(2 338 平方千米),但太湖很浅(最大深度 2.6 米)且高度营养化,全年都有持续的铜绿微囊藻水华暴发。在 2007 年,无锡市的居民因为自来水中有奇怪的味道,以及担心可能从湖水中摄入蓝细菌毒素,转而使用瓶装水长达一个月。人们的这种担忧持续到了今天,他们同时也在努力改善对这一重要水源的水质监控和污染管理。

微囊藻毒素是一系列可溶于水的毒素。许多能引起水华的蓝细菌都能生产这种毒素,但最知名的还是一个遍布世界的物种——铜绿微囊藻。化学上这类毒素被归为多肽,每个分子由一系列的氨基酸以肽键相连而成,如同蛋白质。但不像许多蛋白质那样,多肽不会在沸水中变形,可能因为氨基酸是以一种稳定的环结构排列的(图 34)。这种稳固的特质同样能够抵御细菌类分解者所分泌的能分解蛋白质的酶(蛋白质水解酶)。详

图 34　蓝细菌分泌的有毒多肽微囊藻毒素-LR

细的分析显示，尽管基础的环结构相同，侧边的功能团则多种多样。在富含微囊藻的水中，检测出超过 100 种的微囊藻毒素变体或说"同源物"，其中以微囊藻毒素-LR 毒性最高。

微囊藻毒素对水处理流程的抵御能力掩盖了它们的生物化学活性。一旦进入哺乳动物的体内，这些毒素会在肝脏中被细胞吞噬并阻碍几种关键酶的活性，尤其是磷酸酶。这最终会导致肝脏受损，并给肾脏、脑部和生殖器官带来相应的氧化压力。也有证据证明，微囊藻毒素有致癌性，会干扰微管组装和细胞分裂。饮用含有微囊藻毒素的水会引发恶心、呕吐和消化道疾病，但已知唯一的致死案例发生在 1996 年巴西卡鲁阿鲁的一座医院。院内超过 100 个肾病病人发病，并有 70 人在使用了有微囊藻水华的水库水进行透析后死亡。在医院的水处理系统以及病人的血液和肝脏中也发现有微囊藻毒素。也有许多案例报告，狗和农畜在喝了含有微囊藻等有毒蓝细菌水华的水后死亡。

人们在1950年代首次发现微囊藻的毒效,其时对小鼠的测试发现蓝细菌毒素可以导致小鼠的快速死亡,因而被标为"快速死亡因子"。现在人们知道这是微囊藻毒素。但是在1960年代,加拿大的几头奶牛在饮用含有水华的水死亡后,一种毒性更强、毒发时间更短的蓝细菌毒素被分离了出来,并被标记为"非常快速死亡因子"。最终人们知道这是一种生物碱,把它命名为"鱼腥藻毒素a",它有着强大的神经毒性,能在几分钟内致死。这种蓝细菌毒素首次从富营养湖中常见的固氮种水华长孢藻(鱼腥藻)中分离出来,但人们已经知道蓝细菌有其他几个物种和几个属也能生产这种毒素。

除了微囊藻毒素和鱼腥藻毒素a,引起水华的蓝细菌还能生产一系列对野生动物、家养动物和人类有毒的其他化合物。这当中包括磷酸有机物、生物活性氨基酸和麻痹水母毒素。有些物种能够生产细胞壁物质和其他化合物,并引起皮肤瘙痒和皮肤病,也有许多游泳者在水华污染过的水域游泳导致皮肤过敏的报告。但这当中的某些案例可能由其他原因引起,即感染淡水软体动物和鸭子的幼生扁虫(血吸虫),它也可能钻进人体皮肤并引起"泳者瘙痒"。

净化湖水

一个富营养化的"死湖"能不能恢复到其几近纯净的原始状态?为达成这个重要且野心勃勃的目标,需要对水生植物和藻类,尤其是有毒蓝细菌过度繁殖的机制和过程有所了解。20世纪下半叶,当世界上许多湖泊都在经历人口快速增长和污染物排放增加所带来的影响时，关于营养物的讨论集中在三种元素上：碳、氮和磷。北美洲的肥皂和洗涤剂产业不愿看到他们那些富磷产品的生产受到强制性改变，争辩称碳才是导致富营养化的元素。把富含碳的湖水装在瓶中的短期实验似乎能够支持这个观点，尽管当中某些生物样本给出了模棱两可甚至相互矛盾的结果。

营养物是如何引起富营养化的？哪一种元素扮演着最重要的角色？关于这两个问题的研究，加拿大湖沼学家戴维·W. 辛德勒提供了最有说服力的证据。他的实验是在加拿大的实验湖区（ELA）进行的。加拿大北部这片广阔的花岗岩土地被称为前寒武纪地盾，拥有数以百万计的湖泊和池塘，由最近的冰川活动在岩石上蚀刻而成。这片湖泊聚集的地表景观在位于安大略省北部的一小块区域于1968年被划分出来，用作全湖观测和实验。辛德勒的实验简单易行，其实验结果是整个生态系统给出的，而不是实验室中的人造环境给出的。在一个沙漏状的湖（在ELA目录中编号为226）中，他和他的团队用一张尼龙加固的乙烯幕布横跨切割了湖中央，在西南湖盆施用碳（用的是会被细菌迅速转换为二氧化碳的蔗糖）和氮（硝酸盐）的肥料，而另一边（东北湖盆）在碳氮肥料的基础上加上了磷（磷酸盐）。当中所有比例都和污水处理厂排放的废水接近。

实验结果（图35）十分惊人。在加有碳和氮肥料的一侧，根据光合色素叶绿素的测定结果，藻类生物质几乎没有变化。这一结果在226号湖上表现得尤为有趣。如同其他位于加拿大地盾的湖泊，226号湖溶解无机碳的天然水平很低。假如碳对富营养化有影响，这里是最佳实验场所之一。与之截然相反的是，在

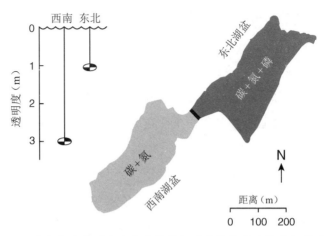

图35 226号湖富碳氮磷东北湖盆的蓝细菌水华暴发及其引起的透明度下降

加有碳氮磷肥料的一侧，有害藻类水华暴发了，主要由固氮蓝细菌为主。水华让水变成绿色并且水质变得浑浊，透明度从约3米下降到1米。除了许多其他水质变量的差异，幕布两侧这种视觉上的对比给政策制定者提供了强有力的证据：磷才是限制水华暴发的关键营养物，保护和修复淡水的工作应该特别关注对点源和非点源排放中磷输入的控制。

在20世纪下半叶，全世界都把关注点放在了对磷的控制上，如寻找磷源、把污水排出湖盆、安装脱磷系统以及规范含磷洗涤剂和其他产品的使用。最早的一个案例是美国华盛顿湖。蓬勃发展的西雅图市以不断增加的速度把污水排放到湖中，在1960年代早期达到每天8 000万升。华盛顿大学的W. 汤米·埃蒙德森在研究中注意到湖水水质下降的先兆，包括营养物的上升和蓝细菌的繁殖。这些发现最终让西雅图市把城市污水计划分流排放到海中。这个计划是历经数年逐步实施的，到1968年再也没有下水道污水排放到湖中。从1964年到1969年这五年时间里,埃蒙德森的团队发现水质有了巨大的改善,夏季藻类浓度下降为原来的1/6,和冬季磷浓度下降的幅度相似。

现在的讨论集中在这个问题上:光控制磷就足够了吗?上层大气和流域内的无机与有机源保证了碳有充足的补充。但是,有几个原因值得把氮也纳入考虑范围。一个反对关注氮的观点是,固氮蓝细菌,如那些在226号湖经历施用碳氮磷肥料后大量出现的物种,有着来自上层大气无穷的气态氮源,以至于无法控制。但是,认为固氮者只从大气中获取部分氮,它们还必须依赖其他氮源,如铵盐、硝酸盐和水中的有机氮,这种说法不完全正确。除此之外,引起湖泊和水库中蓝细菌水华的最令人担忧的物种是微囊藻。这种蓝细菌并不能固定氮,其产生的微囊藻毒素富含氮(见图34;每个分子中有10个氮原子),同时也有证据说明氮的富集能够促进这种毒素的合成。随着土壤排水系统效率的日渐提升,大量富氮富磷的肥料从农用地被排放到水体中。受此刺激,这一有毒物种在全世界范围内似乎都有死灰复燃的态势。

有些湖泊坐落在天然富磷的流域,如新西兰北岛火山高原中部的湖泊以及南美的的喀喀湖,它们和ELA的湖泊有着截然不同的化学成分。对这些湖泊来说,全面的磷管控是不现实的。失衡的磷管控可能也会导致大型植物的扩张,因为这些植物能通过根部获得沉积物中丰富的贮藏磷,同时受益于上层水中的氮富集。最后,淡水湖最终会流入海洋,而滨海环境一般富含磷,氮的浓度有限。假如只考虑移除磷,可能会把富营养化的问题带到下流的这些接收水域中。

出于这些理由,美国和欧盟的环境保护局在污水管控中建议同时移除磷和氮。这个政策决定引来一些争议,因为移除氮的技术昂贵,实操上也比移除磷更困难。针对磷这一单个元素为政策制定者和管理者提供了清晰不含糊的目标,且在这一过程中所有营养物的含量也减少了,处理方式如把处理过的污水通过管道排放到湖盆中(如太浩湖和华盛顿湖),或通过天然或人工湿地同时移除氮和磷。无论当地政策如何,辛德勒从226号湖获得的结果以及在ELA的相关实验强有力地证明了,人类是有能力在一片纯净湖水中迅速引发有害水华的,而管控外部营养供应对保护我们的湖泊免于藻类过度繁殖的危害至关重要。

营养物管控实施后,湖水的恢复并不总是如期望那样快速且彻底的。一部分原因在于"滞后效应":湖水在营养物减少后的恢复阶段的轨迹可能和恶化时的轨迹不一样,特别是在经历持续的有害水华暴发、水体透明度大幅下降的情况后,水下植物群落(如褐藻)会承受损失(图36)。有许多过程会影响这一返

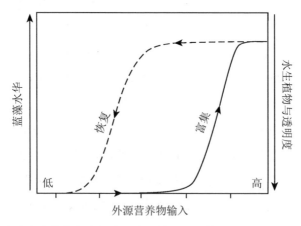

图36 湖泊水华恢复与恶化的迟滞现象

回轨迹，并引起恢复手段的放缓甚至停滞。这部分可能出于生物学原因。比如，在蓝细菌生长多年后，沉积物可能富含孢子和休眠的细胞，成为持续水华的种源。但是，引起恢复放缓最重要的影响来自沉积物中释放出的营养物，尤其是磷，在无氧条件下这种释放会加速（但已知有例外）。这被称为"内部输入"，以便和来自湖泊流域的外部输入区分开来。例如，这种效应对伊利湖的影响就很明显，其严重的有毒水华由外部输入引起，同时混有来自湖泊沉积物的内部磷释放（图19）。释放得越多，意味着藻类的生长就越旺盛，导致更多的生物质被合成以供分解，且氧气消耗增加，加剧了无氧的情况。这种恶性循环很难停止，而目前最好的治理方法就是防止湖水进入氧气耗尽的状态。

湖泊的未来

在关于日内瓦湖的著述中，福雷尔强调了湖泊的物理、化学、生物和人类特点，以及把所有这些方面都纳入考虑以写成一篇综合分析报告的需要，即"一个关于所有详细事实的总览，其中每个专业领域都应由其他领域的研究数据所支持"。在今天，这种纵览全局的视野成了地球系统科学的核心，其中环境的每一方面，从地质物理到人为过程，都应该被视作全球系统中一个互动的部分。拥有一个综合的、系统性的视角看待湖泊，这与处在全球范围气候快速变化背景下的世界淡水资源管理息息相关。

世界上许多地区的湖泊现在都呈现变暖的趋势，平均下来和空气温度升高的速度差不多。但是，变暖的幅度却有很大差别，甚至同一气候区的不同湖泊之间差别也很大，这是因为不同的湖泊有不同的深度、所处风向和透明度。气候变暖往往伴随着极端天气,欧洲和北美洲部分地区的强降水事件已经被认定是有色溶解有机质输入增加的原因,这会导致湖水褐化。这种有机物的富集会改变水生食物网(图16)并降低透明度。这意味着更多的太阳能将被表层湖水吸收,进一步导致水体变暖。

水温的上升会给整个湖泊生态系统带来一系列的影响。蒸发率随温度上升而上升,这可能将湖水平衡推至总体损失的一端,湖面的下降可能会被降雨的改变抵消或加剧。即使水面小型的波动也会对生态系统的重要特征造成严重影响。例如,北美五大湖周边的湿地对候鸟和许多鱼类十分重要,这些半水生栖息地极易受水平衡微小变化的影响。温度的上升也会减小动植物中嗜冷物种所偏好的栖息地的规模,同时会协助来自更温暖气候的物种入侵。

气候变化对湖泊一个稍微没那么明显的影响体现在水的分层上。更暖的条件会让表层水和底层水的温度差异更大,于是密度差异也就更大。分层现象越明显,湖泊对风致混合的抵御能力就越强,因此会削弱氧气在大气和深层水之间的交换,以及营养物在底层水和表层水之间的交换。在坦噶尼喀湖,水温的上升和分层的稳定似乎已经导致真光带的营养供给下降,从而引起浮游植物生产力的下降,以及鱼产量下降30%。这些热分层效应对管理有害蓝细菌水华有着特殊的意义,因为暖水会引发水华,而且在蓝细菌基于气囊和浮力的迁徙中,它们会偏向在稳定的水层活动。

福雷尔描述了日内瓦湖湖盆的大部分居民如何生活在靠近湖岸的地方,以及他们如何成为湖泊生态系统的一部分。后者与一个在当时以及20世纪大部分时间内都受主流认可的观点背道而驰，即人类比自然的地位更高，可以全面统治陆地、大气和水域，以及拥有无限剥削这其中资源的权力，以满足我们不断增长的需求。在福雷尔撰写他著述的第三卷并描述日内瓦湖的人类历史时，住在洛桑的人口约为5.6万，全世界人口在16亿左右，而大气中二氧化碳的含量为296 ppm。而在接下来的100年里，本地和全球的人口都增加为原来的四倍，二氧化碳含量增加了25%。今天，超过80万人从湖中获取水，他们一直关注对营养物和污染物的控制。在世界淡水中日渐常见的新兴污染物包括药物、微塑料（尺径小于5毫米的聚乙烯颗粒）以及经过加工的纳米金属颗粒（1～100纳米）。除此之外，如同许多别处的湖泊，日内瓦湖开始显现气候变化的影响，如底层水变暖、分层和混合现象出现变化，以及某些鱼类产卵日期改变。

人口增长和全球变化对全球范围内湖泊的影响越来越大，这提醒着我们：虽然我们或许是整个生物圈最强大的实体，但我们和我们星球的环境有着密切的互惠关系，并在保护环境完整性以及我们高度依赖的生态系统上有着既定的利益。湖泊是生物多样性的核心；是活泼的"慢河"，水在其中流动、混合和反流；是连接大气和海洋的渠道；是周遭环境的集合体；是过去和现在变化的哨兵。从洪水控制和运输系统，到水、食物和能量的仓库，湖泊是人类社会的关键资源。为保护和维持所有这些价值，需要全球水平的政策决定和行动、湖泊科学和本地管理实践的不断发展，以及对湖沼学标志性的综合研究方法的关注。

索 引

（条目后的数字为原书页码，
见本书边码）

A

A, Lake 湖泊 A 83, 99—103, 105
acid lakes 酸性湖 14, 105—107
acidneutralizing capacity 酸中和能力，见 alkalinity
acidophiles 嗜酸微生物 92
acid rain 酸雨 58
aesthetics 美学 5, 7, 9, 24, 26
Agassiz, Lake 阿加西湖 14
agriculture 农业 11, 14, 21, 25, 97, 110, 112, 115, 124
airshed 空气区 61
algae 藻类 56—57, 78, 114, 121, 126
algal blooms 藻类水华，见 blooms
algal pigments 藻类色素，见 pigments
algal toxins 藻类毒素，见 toxins
alkalinity 碱性 58
allochthonous carbon 异源碳 49, 50, 72
alphine lakes 高山湖泊 85, 92, 97—99
aluminium 铝 58
Amazon Basin dams 亚马孙盆地大坝 111, 113
Amazon floodplain lakes 亚马孙洪泛平原湖泊 20, 51
ammonification 氨化作用 61, 62
ammonium 铵根离子 55, 56, 61, 62, 102—103, 105
amphipods 端足类动物 75, 83, 86

anadromous fish 溯河洄游鱼类 83
anammox 厌氧氨氧化(作用) 61, 62—63
anatoxin-a 鱼腥藻毒素 a 120
anoxia 无氧 38, 51—52, 60, 62, 63—64, 82, 100—101, 115, 126
Antarctic lakes 南极湖泊 91, 92, 93, 97—105
aphotic zone 深水区 29, 71
aquacultrue 渔业 32, 53
Aral Sea 咸海 112—113
archaea 古核生物 56, 59—60, 93, 99, 102, 105
Arctic char 北极红点鲑 83
Arctic floodplain lakes 北极洪泛平原湖泊 97
Arctic lakes 北极湖泊 2, 17, 92, 97—103
ASLO 湖沼学和海洋学学科联盟 7
assimilation 同化作用 61, 62
Aswan dam 阿斯旺大坝 113
autochthonous carbon 同源碳 50

B

bacteria 细菌 48, 49, 52—55, 57, 60—63, 64, 67, 71, 78, 90, 99, 105
Baikal, Lake 贝加尔湖 4, 10, 11, 17, 19, 75, 80, 81—82, 86—87, 94
bathymetry 等深线图 17—18
Belo Monte dam 贝卢蒙蒂大坝 111, 113
benthic boundary layer 底栖边界层 44
benthic zone 底栖区 71
billows 波状运动 44—45
bioamplification 生物放大作用 85

biodiversity 生物多样性 2, 3, 11—12, 48—49, 52—57, 65—69, 74, 94, 101, 113, 117, 123, 129

biofilms 生物膜 74, 95, 98—99

biogeography 生物地理学 97—98

bioprospection 培植 106

bioturbation 生物扰动 23

birds 鸟类 20, 76, 79, 85, 88—89, 92, 94—97, 112, 121, 127

birth of lakes 湖泊的诞生, 见 lake origins

Biwa, lake 琵琶湖 11, 22, 37, 43, 46, 66, 75

Black Sea 黑海 94

black water 黑色湖水 33

blooms 水华 25, 32, 46, 68, 112, 115—126, 128

blue-green algae 蓝绿藻, 见 cyanobacteria

boats 小船 72, 89, 109—110

Bonney, Lake 邦尼湖 99

browning 褐化 127

Buenos Aires, Lake 布宜诺斯艾利斯湖 14

C

Calado, Lago 卡拉多潟湖 20

calcite 石灰石 95

calcium 钙(钙离子) 58, 93

capillary waves 毛细波 38

carbonate 碳酸盐(碳酸根离子) 58, 59, 93, 94, 95

carbon cycle 碳循环 58—60

carbon, inorganic 无机碳 58, 122

carbon, organic 有机碳 50, 54

carbon dioxide 二氧化碳 16, 49—51, 58, 59, 95, 107—108, 128

carbon isotopes 碳同位素 87

carotenoids 胡萝卜素 68, 85

Caspian Sea 里海 89, 93—94, 95

catadromous fish 降河洄游 83

catchment 集水区, 见 watershed

catfish 鲶鱼 11, 51

CDOM 有色溶解有机质 30, 33, 50, 115, 127

Cedar Bog Lake 赛达伯格湖 71

Chad, Lake 乍得湖 20—21

chironomids 摇蚊 73—74, 76, 85

chloride 氯离子 93, 94

cholera 霍乱 55

chlorophyll 叶绿素 68, 122

ciliates 纤毛虫 57

cichlids 丽鱼科 11

Cisco 湖白鲑 82

cladocerans 枝角类动物 24, 66, 76, 77—80, 81, 88

clams 蛤蜊 74—75

climate change 气候变化 16, 21, 22, 90, 92, 98, 104, 110, 114, 126—128

colour of lakes 湖泊颜色, 见 water colour

comammox 全程硝化者 62

Como, Lake (Lago di Como) 科莫湖 12

compensation depth 补偿深度 28

conductivity 电导率 93

connections to sea 与海的连接 2—3, 83, 124, 129

conservation 保护区 25, 90, 94—97

Constance, Lake (Bodensee) 博登湖 11

湖泊

contaminants 污染物 3, 55, 128
convective mixing 对流混合 37
copepods 桡足类动物 11, 66, 76, 77, 80—81, 82, 85, 86—87, 88
Coriolis Effect 科里奥利效应 42—46
Crater Lake 克雷特湖 28, 32
crater lakes, meteoritic 陨坑湖 15—16, 22
crater lakes, volcanic 火山口湖 14—15, 106—108
crayfish 淡水龙虾 65, 76
cryosphere 冰雪圈 98
Crystal Eye of Nunavik 努纳维克的水晶眼 15—16
currents 水流 2, 26, 44—47, 65
cultural values 文化价值 4—5, 15, 21, 24—25, 113, 114
cyanobacteria 蓝细菌 55—56, 61, 68, 95, 98—99, 115—125, 128
cyanotoxins 蓝细菌毒素 118—121, 124
cycles, biogeichemical 生物地化循环 57—65

D

dams 大坝 109—114
Darwin, Charles 查尔斯·达尔文 2, 3
dating of sediments 沉积物的年代鉴定 23
dead lakes 死湖 115, 121
Dead Sea 死海 94
death of lakes 湖泊的死亡 10, 16
decomposition 分解作用 49, 52, 55, 58, 59, 70—71, 117, 118

deepest lake 最深的湖 10
Deep Lake 迪普湖 93
denitrification 反硝化作用 61, 62
density currents 密度流 47
detergents 洗涤剂 121, 123
diatoms 硅藻 11, 23—25, 66, 68, 90, 99
dimictic lakes 二次循环湖 38
dinoflagellates 双鞭毛虫 68
dissimilatry nitrate reducers 异化性硝酸盐还原菌 61
dissolved organic matter 溶解有机质 33, 49, 59
dissolved solids 溶解固体 93
DNA analysis 脱氧核糖核酸分析 24, 52, 66, 69, 75, 83, 101—104, 105, 106
DNA sequence tree 脱氧核糖核酸序列树状图 102
dragonflies 蜻蜓 82, 85
drainage basin 流域，见 watershed
drinking supplies 饮用水供应，见 water

E

ecosystem engineers 生态系统工程师 74
ecosystem services 生态系统服务 1, 7, 8, 25, 66, 109—110, 114, 115, 129
eels 鳗鱼 83
eggs 卵 77, 78—80, 82—83
eicosapentaenoic acid 二十碳五烯酸 85
El'gygytgyn, Lake 埃利格格特根湖 22
Ellsworth, Lake 埃尔斯沃斯湖 15, 104

endemic species 地方特有物种 11, 75, 85, 94, 113
endorheic lakes 内流湖 95—96
English Lake District 英格兰湖区 3, 11—13, 37
EPA (Environmental Protection Agency) 环境保护局 125
EPA (fatty acid) 脂肪酸 88
epilimnion 湖上层 36—37, 41
ephippia 卵鞍 79—80
Erie, Lake 伊利湖 42, 63—64, 90, 118, 126
euglenophytes 裸藻门 32
Eurasian milfoil 凤眼莲 87
eukaryotes 真核生物 56—57
European Union 欧盟 125
eutrophic lakes 富营养湖 32, 51, 62, 63, 64, 74, 115, 121—126
eutrophication 富营养化 30, 110, 114—126
evaporation 蒸发 1, 19, 21, 87, 95, 127
evolution 进化 2, 54, 75, 92
Experimental Lakes Area (ELA) 实验湖区 122—125
exploding lakes 喷发性湖泊 103—105
extinction 灭绝 11, 89, 96, 115
extreme lakes 极端湖泊 91—108
extreme weather events 极端天气事件 127
extremophiles 嗜热微生物 92, 105

F

Fast Death Factor 快速死亡因子 120
fatty acids 脂肪酸 85, 117
FBA 淡水生物协会 12

fish 鱼类 2, 11, 20, 49, 51, 53, 56, 71—77, 82—83, 86—89, 113, 127—128
fisheries 渔业 1, 9, 11, 21, 66, 72, 89, 109, 113, 128
flagellates 鞭毛 49, 57, 67
flaming lakes 产生火焰的湖泊 60
Flathead Lake 弗拉特黑德湖 88—89
flatworms 扁虫 113, 121
flies (dipterans) 蝇类(双翅目) 73, 96
flood control 洪水管控 114
floodplain lakes 洪泛平原湖泊 20, 97
flotation 上浮 116—117, 128
fluorescence microscopy 荧光显微镜 54, 56
flushing time 冲换时间 18—19
food webs 食物网 1, 49, 70—90, 96, 117, 127
Fryxell, Lake 弗里克塞尔湖 99
fungi 真菌 50

G

gastropods 腹足纲 74
gas vesicles 气囊 116, 128
gelbstoff "黄色物质" 33
GenBank 基因银行 101
General Carrera Lake 卡雷拉将军湖 14
Geneva, Lake 日内瓦湖 5—7, 10, 11, 19, 26, 33, 37, 39—40, 45, 47, 65, 66, 70—71, 75—76, 81, 83, 109—111, 115, 128
genomic analysis 基因组分析, 见 DNA
geosmin 土臭味素 118
GIS 地理信息系统 17
glacial lakes 冰川湖 11—14

glass worms 玻璃虫 82

global change 全球变化 3, 8, 10, 16, 90, 92, 104, 127—129

Grande Ethiopian Renaissance Dam 埃塞俄比亚复兴大坝 111

Great Bear Lake 大熊湖 97

Great Salt Lake 大盐湖 95

green algae 绿藻 66, 68, 93, 96

greenhouse gases 温室气体, 见 carbon dioxide, methane

greening 变绿, 见 eutrophication

groundwater 地下水 1

gyres 涡流 46

H

湖泊

Haloarchaea 嗜盐古菌 93

halophiles 嗜盐微生物 92

harmful algal blooms (HABs) 有害藻类水华, 见 blooms

Hauroko, Lake 豪罗科湖 12

highest lake 海拔最高的湖泊 15

high pressure liquid chromatography (HPLC) 高压液相色谱 68

H_2S 硫化氢 57, 101

human health and safety 人类健康和安全 1—2, 20—21, 48, 55, 106—108, 109, 112—113, 118—121

human impacts on lakes 人类对湖泊的影响 9, 11, 21—25, 104, 109—129

humic acids 腐殖酸 33, 50

hydroelectric lakes 发电水库, 见 hydro-reservoirs

hydrogen bonding 氢键 34—35, 38

hydrogen isotopes 氢同位素 87

hydrogen sulfide 硫化氢, 见 H_2S

hydrology 水文学 1, 18—21, 87

hydro-reservoirs 蓄水区 1, 87

hypertrophic lakes 超富营养湖 115

hypolimnion 湖下层 36—37, 112

hyoxia 缺氧 51, 115

hypsographic curve 陆高水深曲线 17

hysteresis 滞后现象 125—126

I

ice 冰 2, 4, 35—37, 51, 60, 91, 92, 98, 100, 103—104

indicator species 指标物种 73, 74

insects 昆虫 20, 24, 73, 82, 86

internal waves 内部波, 见 seiches

invasive species 入侵物种 9, 72, 87—90, 127

inverse stratification 逆向分层 38

inverterd microscopy 倒置显微镜 68

iron 铁 63—64

irrigation 灌溉 112—113

isotope analysis 同位素分析 85—87

Issyk-Kul, Lake 伊塞克湖 94

J

James Bay hydroelectric complex 詹姆斯湾水电设施 111

jellyfish 水母 76

K

kairomones 利它素 82

Kelvin-Helmholtz instabilities 开尔

文-亥姆霍兹不稳定性,见 billows
Kelvin waves 开尔文波 43, 44
kettle lakes 壶穴湖 14, 24
Kinneret, Lake 基尼烈湖 21
Kivu, Lake 基伍湖 130
Kuybyshevskoye 古比雪夫水库 111

L

Lake 226 226 号湖 122—125
lake definitions 湖泊的定义 1, 2, 45, 129
lake level 湖面水位,见 water level
lake origins 湖泊的起源 10—16
lake restoration 湖泊的恢复 121—126
lakes, number in the world 世界上湖泊的数量 4
lake as models 作为模型的湖泊 92, 99
lake as rivers 作为河流的湖泊 1, 2, 20, 129
lake as sentinels 作为哨兵的湖泊 3, 92, 97, 129
land development 土地发展 90, 109
landscape connections 地表景观的连接 1, 2, 7, 33
Langmuir, Irving 欧文·朗缪尔 47
Langmuir spirals 朗缪尔环流 46
largest lake (freshwater) 最大的淡水湖 82
largest lake (saline) 最大的咸水湖 93—94
layering 分层,见 stratification
Léman 莱蒙湖,见 Geneva, Lake
limnetic zone 湖沼区,见 pelagic zone
limnlogy 湖沼学 5, 7, 8, 22, 25, 129

lines on lake surface 湖面的线 47
lipids 脂质 85
littoral zone 沿岸区 65, 71, 74, 77, 87
Llanquihue, Lake 延基韦湖 14
lochs of Scotland 苏格兰的湖 41
lowest lake 海拔最低的湖 94

M

macrophytes 大型植物 50, 59, 65, 74, 77, 87, 90, 114, 117, 121, 124, 126
Maggiore, Lake (Lago Maggiore) 马焦雷湖 12, 37
major ions 主要离子 93
Malawi, Lake 马拉维湖 11
Manicouagan, Lake 曼尼古根湖 15
mats 生物垫,见 biofilms
mayflies 蜉蝣 76
Mead, Lake 米德湖 112
mean depth 平均深度 13, 18
Mendota, Lake 门多塔湖 61
mercury 汞 114
meteorite impacts 小行星撞击,见 crater lakes, meteoritic
meromictic lakes 局部循环湖 99
metals, heavy 重金属 97
methane 甲烷 16, 56, 59—60, 108
methanogenesis 产甲烷作用 59, 60—61, 105
methanotroghy 甲烷营养作用 60—61, 105
methyl isobroneol 甲基异莰醇 118
Michigan, Lake 密歇根湖 12
microbial loop 微生物循环 56

microbiology 微生物学 48—69, 101—106, 116—121

microbiome 微生物群系 48—57, 59, 65, 69, 99, 101, 105

microcosm 微观系统 8, 49—50

microcystins 微囊藻毒素 118—121, 124

microplastics 微塑料 128

midges 蠓虫 73

migration 迁徙 81—83, 89, 94, 113, 117, 128

Mimivirus 拟菌病毒 53

minerals 矿物质 93

mites 螨虫 75

mixing 混合 2, 36—45, 51, 66, 98

mixotrophs 混合营养型 57, 67

molecular techniques 分子技术 48, 52

mollusc 软体动物 74—75

湖泊 Mono Lake 莫诺湖 94—97

monomictic lakes 单循环湖 37

Monoun, Lake 莫瑙恩湖 107

Morar, Loch 莫勒湖 12

morphometry (basin shape) (湖盆) 形态 17—18, 111

mosses 苔藓 99

mud deposition 泥土沉积 39

mussels 贻贝 74, 89—90

N

NALMS 北美五大湖管理协会 31

nanoparticles 纳米颗粒 128

nematodes 线虫 72—73

Nevado Ojos del Salado 奥霍斯-德尔萨拉多火山 15

Ness, Loch 尼斯湖 12

nitrate 硝酸盐(硝酸根离子) 61, 62

nitrification 硝化作用 55, 61, 62, 102—103, 105

nitrite 亚硝酸盐(亚硝酸根离子) 61, 62, 105

nitrate ammonification (DRNA) 硝酸盐氨化作用(异性硝酸盐还原为铵) 61, 62

nitrogen cycle 氮循环 60—63

nitrogen enrichment 氮富集 124—125

nitrogen fixation 固氮作用 61, 122, 124

nitrogen isotopes 氮同位素 86—87

nonpoint sources of nutrients 非点源营养物 115, 122

North American Great Lakes 北美五大湖 4, 14, 39, 59, 63, 75, 80, 89, 127

N:P ratio 氮磷比 84

noxious algae 有害藻类, 见 blooms

number of lakes 湖泊数量 4

nutrient loading 营养物输入 126

nutrients 营养物 41, 45, 65, 66, 98, 112, 113—126, 128

Nyos, Lake 尼奥斯湖 107

O

Ohrid, Lake 奥赫里德湖 11

Ojibway, Lake 欧及布威湖 14

oldest lakes 最古老的湖泊 10, 94

oligochaetes 寡毛虫 73

oligotrophic lakes 贫营养湖 63, 74

Ontario, Lake 安大略湖 43—44

optics 光学 27—33

organic contaminants 有机污染物 85, 97

organic matter 有机质 59
ostracods 介足纲 24, 76
overfertilizion 营养物过多 114
overturn 湖水对流 37
Owens Lake 欧文斯湖 96
oxycline 氧跃层 100—103
oxidation 氧化(氧化作用) 57, 62, 63—64
oxygen 氧气 36—37, 41, 44, 45, 47, 50, 51—52, 62, 74, 100—101, 110, 115, 117, 128
oxygen isotopes 氧同位素 87

P

palaeolimnology 古湖沼学 1, 15, 21—25
paradox of the plankton 浮游生物悖论 48, 69
parasites 寄生虫 53, 57, 66, 121
parthenogenesis 孤雌生殖 78—79
pathogens 病原体 9, 55
pelagic zone 浮游区 65, 71—72
periphytn 底栖植物 50, 74
permafrost 永久冻土 16, 60, 105
pesticides 杀虫剂 85, 97
pH 酸碱值 58, 92, 105—107
phages 噬菌体 53
phantom midge 幽灵蚊 82
phramaceuticals 药品 128
phosphorus 磷 23—25, 63—65, 84, 121—126
photic zone 真光带 28—29, 65, 66, 71
photochemistry 光化学 49, 51
photodamage 光损害 68
photosynthesis 光合作用 28, 37, 48—51, 56—57, 59, 65, 98, 99, 101, 117
photosynthetic sulphur bacteria 光合硫细菌 101
phytoflagellates 鞭毛植物 67, 93
phytplankton 浮游植物 48, 49, 50, 59, 65—69, 74, 84, 86, 88, 117
picocyanobacteria 微蓝细菌 55—56, 68
picoeukaryotes 微真核生物 66
pigments 色素 24, 32, 56, 65, 67, 68, 77—78, 85, 98, 99
Pingualuk Lake 平圭勒湖 15—16
plankton 浮游生物；参见 paradox of the plankton
plants 植物，见 macrophytes
Poincaré waves 庞加莱波 43—44
point sources of nutrients 营养物点源 115, 122
polar lakes 极地湖泊，见 Antarctic lakes; Arctic lakes
pollen 花粉 23—25
pollution 污染 22, 46, 47, 58, 65, 73, 85, 97, 119, 121, 128
ponds 池塘 2, 3, 4, 7, 15—16, 32, 77, 78, 80, 82, 87
population growth (human) 人口增长 1, 6, 96, 114, 121, 123, 128
precipitation 析出 87
primary production 初级生产 28, 45, 47, 48—51, 65, 68, 84, 90, 98, 99
profundal zone 深水区 66, 70—71, 74, 81
proglacial lakes 冰前湖 14
Proteobacteria 变形菌 54—55
protists 原生生物 49, 56—57, 81, 90, 98, 99
protozoa 原生动物，见 protists
psychrophiles 嗜寒微生物 92, 98
psychrotolerance 耐寒微生物 98

PUFA 多不饱和脂肪酸 85

R

rare biosphere 稀有生物圈 69
ratio, Redfield 雷德菲尔德比值 84
Redon, Lake 勒东湖 97
reduction (chemical) 还原(还原作用) 57, 62, 63—64
regime shift 水文情势的转变 90
respiration 呼吸作用 28, 59, 117
reproduction 繁殖 79, 85
resevoirs 水库 36, 109—114, 120
Rinihue Lake 里尼韦湖 14, 20
ripples 涟漪 38
rivers and dams 河流和大坝，见 dams
RNA 核糖核酸 52, 84, 101
rotifers 轮虫 66, 75, 76—77, 79, 81
Ruaehu crater lake 鲁阿佩胡火山湖 106—107
Rybinskoye 雷宾斯克水库 111

S

St-Charles, Lake (lac Saint-Charles) 圣查尔斯湖 19, 36—37, 87, 111—112
saline lakes (salt water lakes) 碱水湖(咸水湖) 91—97, 112
salinity 盐度 19, 91, 93, 100
salmon 鲑鱼 88—89
saltiest lakes 最咸的湖泊 94
satellite remote sensing 卫星遥感测量 4, 31
satellite tracking 卫星追踪 83

schisosomiasis 血吸虫病 113, 121
scuds "飞毛腿" 75
sculpins 杜父鱼 11, 86
Sea of Galilee 加利利海，见 Kinneret, Lake
seals 海豹 11, 86—87, 94
seasons of a lake 湖泊的季节 36—38, 66
seawater 海水 93, 100
Secchi depth 西奇深度(透明度) 27—28, 115, 122—123, 126
secondary compounds 次级化合物 117—118
secondary production 次级生产 84
sediments 沉积物 10, 13, 15, 21—22, 39, 47, 60, 63—65, 70—71, 73, 82, 89, 105, 124, 126
sedimentation 沉积作用 10, 39, 66, 71, 114, 117
seiche, internal 内部假潮 41—44
seiche, surface 表面假潮 26, 39—42
shrimps 虾 75, 76, 81, 88—89, 96, 115
SIL 国际湖沼学学会 7, 74
silica 硅玻璃壳 23, 66
size of lakes 湖泊的大小 4, 15—16
sludge worms 污泥蠕虫 73
small lakes 小型湖泊 15—16
snails 腹足类动物 11, 74, 76
snakes 蛇 20
snow 雪 2, 37, 92, 98, 100, 101
Snowball Earth 雪球地球 99, 105
solubility, carbonate 碳酸盐溶解度 59
solubility, oxygen 氧气溶解度 52
solute 溶质 8, 93

sponges 海绵 11
stable isotopes 稳定同位素，见 isotope analysis
stoichiometry, ecological 生态化学计量学 84
stratification 分层现象 2, 26, 36—38, 99—101, 115, 127—128
stress effects 压力影响 79
stromatolites 叠层石 99
sturgeon 鲟鱼 113
subglacial lakes 冰下湖 15, 103—105
sulfate 硫酸盐（硫酸根离子）57—58, 93, 94
sulfide 硫化物 57, 64, 101, 118
sulfur 硫 101
sulfur isotopes 硫同位素 87
sunlight 阳光 26—33, 49, 51, 66, 98, 101, 103, 117
Superior, Lake 苏必利尔湖 12, 82
swimmer's itch 泳者瘙痒 121

T

Tahoe, Lake 太浩湖 10, 31, 37, 42, 81, 111, 125
Taihu, Lake 太湖 118—119
Tanganyika, Lake 坦噶尼亚湖 10, 11, 128
Taupo, Lake 陶波湖 15, 37
tectnic basins 构造湖盆 10, 94, 103
temperature 温度 35—38, 42—44, 52, 68, 79, 91, 92, 93, 100—101, 105, 106, 127—128
thaw lakes 解冻湖 16, 60, 77
thermokarst lakes 热喀斯特湖，见 thaw lakes

thermocline 温跃层 36, 37, 41, 43, 45, 81
Three Gorges River dam 三峡大坝 111, 112, 113, 114
Tibetan Plateau 青藏高原 94
Titicaca, Lake (Lago Titicaca) 的的喀喀湖 5, 11, 18—19, 37, 124
top-down effects 自上而下效应 88—89
toxic lakes 有毒的湖泊 118—121, 124
toxins 毒素 118—121, 124
transfer function 传递函数 23, 25
transparency 透明度，见 water clarity
transport 运输 9, 90, 109—110, 113, 115
trophic cascade 营养级联 88—89
trophic-dynamic concept 营养动力学概念 71
trophic state 营养状态 74
tropical floodplain lakes 热带洪泛平原湖泊 20, 51, 113
trout 鳟鱼 37, 89
turbulence 扰动，见 mixing

U

ultramicrobacteria 超微细菌 54
unionid clams 珠蚌 75, 90
Untersee, Lake 温特塞湖 98
upwelling 上涌 42
Urmia, Lake 奥卢米耶湖 112
UV-radiation 紫外线辐射 31, 33, 77—78, 85, 98

V

Vanda, Lake 万达湖 32, 91, 97, 99, 101

索引

varzéa 低洼地 20
veligers 面盘幼体 90
Very Fast Death Factor 非常快速致死因子 120
Victoria, Lake 维多利亚湖 11, 87
viral shunt 病毒分流 53
viruses 病毒 48, 49, 53—54, 56, 69, 99
volcanoes 火山, 见 crater lakes, volcanic
Vostok, Lake 沃斯托克湖 15, 103—104

W

Wakatipu, Lake 瓦卡蒂普湖 12
Walden Pond 瓦尔登湖 3, 24—25
Washington, Lake 华盛顿湖 123, 125
Wastwater 沃斯特湖 12, 13, 111
water 水
    balance 水平衡 1, 18—21, 87, 127
    clarity 水的澄清度 27—31, 90, 114—115, 127
    colour 水的颜色 31—33, 127
    conductivity 水的电导率 93
    density 水的密度 34—35, 38, 91
    dielectric constant 水的介电常数 92
    drinking 饮用水 1, 6, 9, 36, 55, 87, 90, 96, 109, 117—121
    freezing 水的冻结 4, 35, 37
    level 水位 9, 14, 18—21, 40—41, 95—97, 109, 112, 113, 127
    molecules 水分子 34—35
    odour 水的臭味 57, 101, 118, 119

optical properties 水的光学性质 30
pollution 水污染 46, 47, 58, 65
plants 水生植物, 见 macrophytes
quality 水质 11, 31, 55, 65, 68, 73, 112, 122, 123
residence time 水的滞留时间 18—19, 112
salinity 水的盐度 91, 93
taste 水的味道 117—119
toxins 水的毒素 118—121, 124
weeds 水草, 见 macrophytes
watershed 流域 2, 58, 61, 64, 72, 111, 124
waves 波浪 26, 38—45
Western Brook Pond 西布鲁克池 4
wetlands 湿地 3, 113, 127
Whillans, Lake 惠兰斯湖 15, 104—105
whitecaps 白浪 39, 47
whitefish 淡水白鱼 66, 83
whiteings 白垩化 59
wind and currents 风和水流 26, 45—47
wind mixing 风致混合 2, 43, 91, 127
Windermere, Lake 温德米尔湖 12, 13, 19, 63, 66, 111
winter limnlogy 冬季湖沼学 36—38, 60, 85, 98
worms 蠕虫 70, 72—73

Z

zooplankton 浮游动物 16, 49, 56, 66, 67, 69, 72, 76—82, 86, 87, 90, 117

Warwick F. Vincent

LAKES

A Very Short Introduction

Preface

In August 2018, I had the good fortune to visit Nanjing for the 34th Congress of the International Limnological Society (SIL). This society for the science of inland waters was founded in 1922, and since that time has held conferences every two to three years in different parts of the world. This was only the second time that SIL had been held in Asia, and it was the first time in China.

The Nanjing meeting was located in the beautiful Jianye district of wide avenues, gleaming hotels and glass office towers, some still in construction and growing higher every night, it seemed to me at the time. The hospitality by our Chinese hosts was outstanding, and it was a great pleasure to meet so many students and researchers from China, and from around the world.

This was also a time to celebrate my former PhD advisor from the University of California at Davis, the distinguished professor of limnology Dr. Charles R. Goldman, who was turning 88 years old that year. We ran a special symposium as a tribute to Professor Goldman entitled "Global Lessons from Lakes of the World" and led by my colleague Professor Michio Kumagai, former President of the Japanese Society of Limnology. This was followed by a memorable dinner of Chinese delicacies with Professor Goldman in the company of many of his friends, colleagues and former students, and hosted by Professor Peimin Pu, renowned Chinese limnologist and former director of the Taihu Laboratory for Lake Ecosystem Research.

At the end of that week, I visited Lake Taihu on a SIL field trip led by Professor Hans Paerl (also a former student of Professor Goldman). We learned about the critical importance of this vast water supply for more than 40 million people, and the wide range of research in progress to understand and manage its water quality problems, now with the additional challenge of global climate change.

This experience from my Nanjing visit made me very interested in the possibility of a Chinese edition of the short introduction to lakes that I had published with Oxford University Press in 2018, and that was subsequently published in French in Quebec City and Paris. I am especially pleased that the publisher for this Chinese edition, Yilin Press, is based in Nanjing, which played such an important role in my introduction to limnology in China. I thank Ms. Dan Xu and her colleagues at Yilin Press for their encouragement and valuable assistance throughout this project. I am particularly grateful to Mr. Tianyang 'Dirk' Deng, who bravely agreed to translate the English version into Chinese while he was a PhD student in chemistry with us here at Laval University. As his thesis co-director, I was always worried that this would take time away from his exciting research to develop environmental analytic systems, combining microfluidics with advanced spectroscopy, but fortunately, he was able to balance these tasks, and I am especially pleased that he became very knowledgeable about lake science in the process.

I thank the many limnologists who provided expert feedback and review comments during and after the publication of the English edition, including B. Beisner, S. Bonilla, R. Cory, A. I. Culley, G. W. Kling, M. Kumagai, U. Lemmin, I. Laurion, C. Lovejoy, S. MacIntyre, S. Markager, F. Pick, R. Pienitz, M. Rautio, G. Schladow, R.W. Sterner, P. Vanrolleghem and A. Vigneron. I also thank Amanda Toperoff for the superb graphics, and the Canadian funding sources for my research on lakes, including NSERC, FRQNT, CRC, CFI, NCE-ArcticNet and CFREF-Sentinel North.

Finally, I thank again the organizers of the SIL congress in Nanjing for their warm welcome to China, and for the wonderful introduction that they provided to China's diverse lake and river ecosystems. I hope that this book will help capture the imagination of students in the environmental sciences, and that it will encourage all readers to learn more about the mysteries that lie beneath the surface of the lakes of the world.

Warwick F. Vincent

Laval University, Quebec City, Canada

In memory of
Dennis A. Walter
(1938–2013)

Contents

Acknowledgements i

List of illustrations iii

1 Introduction 1

2 Deep waters 6

3 Sunlight and motion 26

4 Life support systems 48

5 Food chains to fish 70

6 Extreme lakes 91

7 Lakes and us 109

Further reading 131

Acknowledgements

The author would like to thank: François D. C. Forel for permission to include material from the publications of his great grandfather, François A. Forel; Amanda Toperoff for the superb graphics; Beatrix Beisner, Sylvia Bonilla, Rose Cory, Alexander Culley, George Kling, Michio Kumagai, Ulrich Lemmin, Isabelle Laurion, Connie Lovejoy, Sally MacIntyre, Stiig Markager, Frances Pick, Reinhard Pienitz, Milla Rautio, Geoffrey Schladow, Peter Vanrolleghem, and Adrien Vigneron for their valuable feedback on sections of the manuscript; Latha Menon and Jenny Nugee at Oxford University Press for their excellent editorial support; and the granting agencies that have financially supported the author's research on lakes, notably the Natural Sciences and Engineering Research Council of Canada and the Fonds de recherche du Québec—Nature et technologies.

List of illustrations

1 Lakes as sentinels, integrators, and conduits **3**

2 Interactions that affect lake ecosystem services in the changing global environment **8**

3 Lakes of the English Lake District and their catchments **12**
Adapted and redrawn from A. E. Ramsbottom (1976), 'Depth Charts of the Cumbrian Lakes', *Scientific Publications of the Freshwater Biological Association* (FBA) 33: 1–39, by permission of the FBA.

4 Pingualuk Lake, the 'Crystal Eye' of northern Quebec **16**
Photograph by D. Sarrazin (CEN), reproduced by permission.

5 Bathymetry and the area-depth curve for Lake Baikal, Russia **17**
Based on data from The INTAS Project 99-1669 Team (2002), 'A New Map of Lake Baikal'.

6 Differences in water residence time among lakes **19**

7 The shrinking of Lake Chad in central Africa caused by ongoing dry conditions **21**
Adapted from multiple sources, including the United Nations Environment Programme.

8 Environmental changes over the last 300 years recorded in the lake sediments of Walden Pond, USA **24**
Replotted with permission from the data in D. Köster et al. (2005), 'Paleolimnological Assessment of Human-Induced Impacts on Walden Pond (Massachusetts, USA) using Diatoms and Stable Isotopes', *Aquatic Ecosystem Health & Management* 8: 117–31.

9 Penetration of sunlight into a lake measured with an underwater light meter **29**

10 Hydrogen bonding and the strange density–temperature relationship for water **35**

11 Seasonal changes in Lake St-Charles, the water reservoir for Quebec City **36**
Based on data in S. Bourget, 'Limnologie et charge en phosphore d'un réservoir d'eau potable sujet à des fleurs d'eau de cyanobactéries: le lac Saint-Charles, Québec', MSc Thesis, Université Laval, Québec, Canada (2011), with permission.

12 The surface seiche is caused by the wind pushing water to one end of the lake **41**

13 Poincaré waves on the thermocline of Lake Ontario **43**
Redrawn from F. M. Boyce (2011), 'Some Aspects of Great Lakes Physics of Importance to Biological and Chemical Processes', *Journal of the Fisheries Research Board of Canada* 31: 689–730, reproduced by permission of Canadian Science Publishing. © Canadian Science Publishing or its licensors.

14 Billows across the thermocline **44**

15 The gyres of Lake Biwa, Japan **46**
Reproduced by permission of M. Kumagai (Ritsumeikan University).

16 From sunlight to diverse microbes and the aquatic food web **49**

17 The aquatic carbon cycle **59**

18 The aquatic nitrogen cycle **61**
Reprinted from J.-É. Tremblay et al. (2015), 'Global and Regional Drivers of Nutrient Supply, Primary Production and CO_2 Drawdown in the Changing Arctic Ocean', *Progress in Oceanography* 139: 171–96, with permission from Elsevier.

19 Phosphate release from Lake Erie sediments **64**
Replotted by permission from the data in X. Ding et al. (2015), 'Characterization and Evaluation of Phosphate Microsensors to Monitor Internal Phosphorus Loading in Lake Erie Sediments', *Journal of Environmental Management* 160: 193–200.

20 The colonial phytoflagellate, *Dinobryon divergens* **67**
Photomicrograph provided by I. Fournier (Université Laval) and J. D. Wehr (Fordham University), reproduced by permission.

21 The ecological zones of a lake **71**

22 The blind shrimp (*Niphargus forelii*) of Lake Geneva **76**
Reproduced by permission from: F. A. Forel (1904), *Le Léman: Monographie limnologique* (Lausanne: F. Rouge), vol. III, Fig. 180.

23 Photomicrograph of the zooplankton species *Daphnia umbra* from a lake in Finland **78**
Photomicrograph by P. Junttila, reproduced by permission.

24 Asexual and sexual reproduction by cladoceran zooplankton **79**
Adapted from A. J. Horne and C. R. Goldman, *Limnology* (Columbus: McGraw-Hill, 1994), by permission.

25 Photomicrograph of the copepod *Aglaodiaptomus leptopus* **80**

Photomicrograph by T. Schneider (Université du Québec à Chicoutimi), reproduced by permission.

26 The pelagic food web in Lake Baikal **86**

Based on the data in K. Yoshii et al. (1999), 'Stable Isotope Analyses of the Pelagic Food Web in Lake Baikal', *Limnology and Oceanography* 44: 502–11.

27 Food web changes at Flathead Lake after the invasion of mysid shrimps **88**

Redrawn with permission from the data in: B. K. Ellis et al. (2011), 'Long-Term Effects of a Trophic Cascade in a Large Lake Ecosystem', *Proceedings of the National Academy of Sciences U.S.A.* 108: 1070–5.

28 Tufa towers at Mono Lake, California **95**

Photograph from the Mono Lake Committee, reproduced by permission.

29 Layers of water of different salinities, temperatures, and oxygen **100**

Based on data published in NEIGE (2016), 'Water Column Physico-Chemical Profiles of Lakes and Fiords along the Northern Coastline of Ellesmere Island, v. 1.0', *Nordicana*, D27, doi:10.5885/45445CE-7B8194DB81754841.

30 Tree showing the genetic relatedness of three archaeon strains in Lake A **102**

Replotted from the data in J. Pouliot, P. E. Galand, C. Lovejoy, and W. F. Vincent (2009), 'Vertical Structure of Archaeal Communities and the Distribution of Ammonia Monooxygenase A Gene Variants in Two Meromictic High Arctic Lakes', *Environmental Microbiology* 11: 687–99.

31 Highly acidic lake **107**

32 A traditional merchant vessel **110**

Reproduced by permission from: F. A. Forel (1904), *Le Léman: Monographie limnologique* (Lausanne: F. Rouge), vol. III, fig. 233.

33 Photomicrograph of the toxic bloom-former *Microcystis aeruginosa* **116**

34 The toxic peptide microcystin-LR produced by cyanobacteria **119**

35 Cyanobacterial bloom development **123**

Based on information in P. S. S. Chang et al. (1984), 'Zooplankton in Lake 226: Experimental Lakes Area, Northwestern Ontario, 1971–1978. Data', *Canadian Data Report of Fisheries and Aquatic Sciences* 484: 1–208.

36 Hysteresis in the recovery versus degradation of a lake **126**

Based on M. Scheffer and S. R. Carpenter (2003), 'Catastrophic Regime Shifts in Ecosystems: Linking Theory to Observation', *Trends in Ecology & Evolution* 18: 648–56.

Chapter 1
Introduction

What is a lake? At first glance, this seems like such an easy question: a lake is simply a body of water surrounded by land. But this sterile, physical definition is only a beginning, and there are so many other more interesting ways to consider the nature and meaning of lakes. For freshwater biologists, a lake is an oasis in the landscape where microbes, plants, and animals form networks of interaction, and where species, food webs, and ecological processes await discovery. Many environmental scientists think of lakes in more chemical terms, as living reactors that exchange gases with the atmosphere. These are places that collect and transform materials washed in from the surrounding catchment, and where aquatic plants and algae produce new organic matter by photosynthesis. Some of my colleagues study microscopic fossils that occur in lake sediments, and for these researchers, lakes are rich storehouses of information that can tell us about the past, inform our present, and help guide our plans for the future.

For water engineers and society, lakes are essential resources to be managed, modified, even created, to address the ever increasing demand for drinking water, hydropower, fish production, and other ecosystem services. To maintain these services requires close attention to the balance of surface and groundwater inflows, evaporation, extractions, and outflows that together govern the amount of water remaining in the lake basin. Water is in terribly short supply in many parts of the world, and this balance of gains and losses is becoming ever more precarious and challenging to manage in our changing global climate.

In physical terms, a lake is a body of water that is constantly in motion, energized by the sun and the wind. Depending on season, the lake may be composed of layers that differ, sometimes surprisingly, in temperature, oxygen, colour, salt content, and many other properties, with periods of mixing each year that break down this layered structure. Lakes are connected, slow-moving conduits through the landscape: water moves from the inflows to the outflow of the lake, but this orderly, riverine flow path is continuously disrupted by wind-induced swirls, gyres, and counter-currents, while waves form and break, even out of view beneath the surface.

When I fly over the Canadian North to my field sites each summer, the landscape passes below as archipelagos of glittering freshwaters or, at more northerly, cooler latitudes, as snow-capped plates of lake ice set into the undulating tundra. In some of our work, we have been interested in the dispersal of microscopic life among these constellations of Arctic lakes and ponds, and how the individual waterbodies then select their ensemble of species. For those who study the evolution of fish and other aquatic life, the oldest lakes are island laboratories, where the processes of colonization, genetic shifts, and speciation can help us understand how the biodiversity of our planet has evolved, and how it is continuing to change. Charles Darwin even speculated about how life might have first arisen 'in some warm little pond with all sorts of ammonia and phosphoric salts'.

Lakes are the lowest points in the landscape, before they discharge ultimately (with some exceptions) to the ocean. In this way they can be thought of as integrators of their surroundings (Figure 1), reflecting the combined effects of water supply from their drainage basin (also called catchment or watershed), vegetation, geology, and the natural and human history of their environment. Lakes are

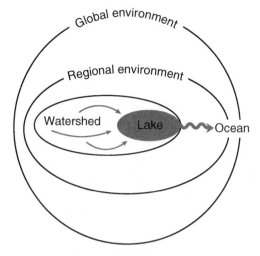

1. Lakes as sentinels, integrators, and conduits.

therefore indicators of environmental change, and can be viewed as sentinels of not only the current magnitude and legacy effects of local human activities, but also as regional and global sentinels of climate change, contaminant dispersal, and biodiversity shifts that are taking place throughout the world.

Then there is the question of size. Could Darwin's warm little pond be considered a lake? Some authors define ponds to be bodies of water of a depth that can be waded across, although in boggy wetlands, such a test would be ill advised. Another definition is that ponds freeze to the bottom while lakes do not, but this especially Canadian view of the aquatic world will not always be very helpful elsewhere; besides, ice-covered bottom waters are impressively resistant to freezing, even in Canada. Visitors to the English Lake District will find that the smaller bodies of water are called 'tarns', and larger ones are called 'lakes', 'meres', or 'waters', with no clear consensus on any of these terms. One of the most famous literary lakes in North America is called Walden Pond, and to further confuse this quagmire of terminology, the people of Newfoundland call most of their lakes 'ponds'—including Western Brook Pond that is 16 kilometres (km) long and 165 metres (m) deep. So all in all, it's best to consider lakes and ponds together, and to apply the word 'lakes' as a general term that encompasses the full range of waterbodies.

Size does become important if we want to know how many lakes there are in the world, in a particular country, or in our local surroundings. We need to set a cut-off value of minimum area, and with improvements in satellite remote sensing, and now increasingly with drones, the imaging threshold becomes smaller each year, as does the lower bound for lake inventories. High resolution satellites can readily detect lakes above 0.002 kilometres square (km^2) in area; that's equivalent to a circular waterbody some 50m across. Using this criterion, researchers estimate from satellite images that the world contains 117 million lakes, with a total surface area amounting to 5 million km^2. In Canada, we like to think that we have the largest number of lakes of any nation, including the North American Great Lakes that we share with the USA and that hold 20 per cent of the world's surface freshwater. But the world map on my office wall reminds me that Russia is also a large, lake-rich country, and that one of its lakes, Lake Baikal, is the deepest in the world and holds another 20 per cent of the surface freshwater on Earth.

The aim of this book is to provide a condensed overview of scientific knowledge about lakes, their functioning as ecosystems that we are part of and depend upon, and their responses to environmental change. Of course there are other, non-scientific, but equally varied reasons that lakes are important to human inquiry and culture. The mysterious, unsettling nature of deep or black waters has long held the interest of novelists and poets, for example Sylvia Plath in her 'Crossing the Water', and William Wordsworth's 'The Prelude', about his midnight traverse as a child across a dark troubling lake. The presence of mythical creatures that inhabit such depths, such as the Taniwha in Maori legends about New Zealand lakes and seas, has been the subject of oral histories in many parts in the world, and lakes may be of wider spiritual significance. In Bolivia and Peru, an ancient legend attributes the origin of the Incan civilization to the births of Manco Cápac and Mama Ocllo, brought up by the sun god Inti from the depths of Lake Titicaca. Lakes as mirrors and as multi-hued palettes of colour have captured the imaginations of artists, musicians, and writers of many cultures, and today attract multitudes of visitors to lakeshores each year. Some writing, including the classic haiku of Matsuo Basho, evokes the audible dimension of lakes and ponds ('the sound of water'), and in his monograph on dreams, Gaston Bachelard considers how the water of pools, ponds, fountains, lakes, and streams is a primary element of 'material imagination' and reverie.

This volume focuses on the science—and my hope is that this short introduction will allow the reader to view the next lake visited with a greater sense of wonder and a desire to learn more about the remarkable features that lie at and beneath its surface. Each chapter briefly introduces concepts about the physical, chemical, and biological nature of lakes, with emphasis on how these aspects are connected, the relationships with human needs and impacts, and the implications of our changing global environment. Lake science has a long history of observation and discovery, and many excellent textbooks are available that describe lake ecology in scholarly depth. These books encompass a broad sweep of established theory and new information, but the roots of much of this knowledge can be traced back to a career decision by a young scientist in the 19th century, at beautiful Lake Geneva, on the edge of the Swiss Alps.

Chapter 2
Deep waters

> I soon put to myself two options: create my own research laboratory in anatomy, histology and physiology, the subjects that I had to teach at the Faculty of Science... Or take for my laboratory and my aquarium this lake that offered me its mysteries and beckoned me to study them. My choice was soon made...
>
> F. A. Forel

Returning to the shores of Lake Geneva as a newly trained scientist and medical doctor, François A. Forel decided upon a career path that would guide his many decades of research, and that would ultimately lay the foundation for modern lake science. Forel was born and grew up on the Swiss side of the lake that lies on the border of Switzerland and France, and he was keenly aware of the wide-ranging importance of the lake to the people who lived around it. First and foremost, Lake Geneva was the primary source of drinking water for its lakeshore communities, including the rapidly growing city of Lausanne where Forel was appointed to teach at the Academy (today, the University of Lausanne).

Forel's father had captured his imagination as a child by bringing him out onto Lake Geneva, known in French as 'le Léman', to explore the ancient stilt-house villages found at several places in its inshore waters. The ruins of these Bronze Age settlements lay beneath the surface, and the archaeological artefacts found at these sites attested to the longstanding relationship between humans and the lake. Forel was also aware of the great commercial value of Lake Geneva for fisheries and transport, and he later quantified these resources in economic terms. He also appreciated the aesthetic appeal of the lake and its mountainous surroundings, and he enjoyed the company of landscape artist François L. D. Bocion. But he was especially taken by the idea that Lake Geneva's deep, blue waters held scientific mysteries and secrets, many of which could eventually be revealed by careful study.

Forel had returned to Lausanne and the family home in nearby Morges in 1867, at the age of 26, after some eleven years of schooling and medical training in Switzerland, France, and Germany. His decision to study 'all aspects of the lake' was initially greeted with some concern by his former professor in Germany, who advised the young man to adopt a more focused and specialized approach. Not to be dissuaded, Forel launched his research on a myriad subjects, from waves and currents, the penetration of sunlight, and the nature of chemicals in the water, to studies on the plants, animals, and microbes that live throughout the lake. It was not until many years later, however, that he brought all of these disparate studies together into an overall synthesis.

In the preface to volume one of his comprehensive monograph on Lake Geneva, published in 1892, Forel coined the word 'limnology', from the Greek 'limne' for lake, and he defined this new integrative science as 'the oceanography of lakes'. Today, limnology has been extended to include rivers, wetlands, and even estuaries, but its primary focus remains lakes and ponds. There are many professional societies around the world for researchers in this field, notably the International Society of Limnology (SIL) and the Association for the Sciences of Limnology and Oceanography (ASLO). The word limnology, however, is a scientific term that is not well known outside the field. The word 'lake' that we use in English for bodies of freshwater is derived not from a familiar root (like Greek 'okeanos' in oceanography) but from the Latin word 'lacus' meaning basin. On the other hand, the concept of limnology is intuitively straightforward and appealing, and is highly relevant to our present-day goals of lake protection, restoration, and management.

For Forel, the science of lakes could be subdivided into different disciplines and subjects, all of which continue to occupy the attention of freshwater scientists today (Figure 2). First, the physical environment of a lake includes its geological origins and setting, the water balance and exchange of heat with the atmosphere, as well as the penetration of light, the changes in temperature with depth, and the waves, currents, and mixing processes that collectively determine the movement of water. Second, the chemical environment is important because lake waters contain a great variety of dissolved materials ('solutes') and particles that play essential roles in the functioning of the ecosystem. Third, the biological features of a lake include not only the individual species of plants, microbes, and animals, but also

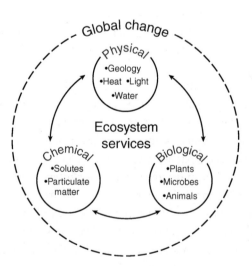

2. Interactions that affect lake ecosystem services in the changing global environment.

their organization into food webs, and the distribution and functioning of these communities across the bottom of the lake and in the overlying water.

There were two aspects of Forel's new science that set his thinking apart from many of his contemporaries, and even from that of many specialists subsequently. As he stated in his earliest definition of limnology, studies of a lake from different points of view can inform each other and show connections, producing an integrated picture of the ecosystem. The interactions among the physical, chemical, and biological properties of the lake were of special interest. For example, he showed how terrestrial organic materials could give rise to the greenish hues of the lake and affect the clarity of the water, how the weathering of rocks in the surrounding catchment affected the mineral chemistry of the lake, and how bottom-dwelling animals were connected to the life and death of plankton in the overlying waters. In contrast to the popular idea of a lake as an enclosed microcosm, he wrote in 1891 that:

> Rather, it communicates with the rest of the world, via its constant exchange of gases with the overlying atmosphere, via its outflow of water carrying away dissolved and non-dissolved substances, and via its tributaries that deliver new materials into the lake.

The second aspect concerned human beings. Forel recognized that lakeshore residents were actually part of the Lake Geneva ecosystem, and that they depended on the lake in many ways to provide them with services such as safe drinking water, the commercial fishery, a waterway for transport of people and cargo, and the aesthetic pleasure and mental well-being of living close to water. All of these benefits can become less available as a result of human impacts on the environment, and Forel observed many examples in Lake Geneva, including the defective regulation of water level, the introduction of invasive aquatic species, and contamination of the lake with human pathogens from sewage. This idea that people are a component part of the ecosystem is a concept that was not fully recognized through much of the 20th century. Yet it is crucially important today as we confront the increasing impacts of global change, and the challenges of sustaining the biosphere that we are both part of and dependent upon.

Birth and death of lakes

Forel devoted many pages of the first of his three-volume treatise on Lake Geneva to considering the possible origins of its basin, and he also described the way that sediments were accumulating in the lake, especially those derived from the glacier-fed, upper Rhône River with its milky suspension of mineral particles. This continuous accumulation of materials on the lake floor, both from inflows and from the production of organic matter within the lake, means that lakes are ephemeral features of the landscape, and from the moment of their creation onwards, they begin to fill in and gradually disappear. The world's deepest and most ancient freshwater ecosystem, Lake Baikal in Russia (Siberia), is a compelling example: it has a maximum depth of 1,642m, but its waters overlie a much deeper basin that over the twenty-five million years of its geological history has become filled with some 7,000m of sediments.

Lakes are created in a great variety of ways: tectonic basins formed by movements in the Earth's crust, the scouring and residual ice effects of glaciers, as well as fluvial, volcanic, riverine, meteorite impacts, and many other processes, including human construction of ponds and reservoirs. Tectonic basins may result from a single fault, as in Lake Baikal and Lake Tanganyika (eastern Africa), or from a series of intersecting fault lines. Lake Tahoe (USA), for example, was created by block faulting giving rise to its rectangular, horse-trough shaped basin with a great average (300m) as well as maximum (501m) depth. The oldest and deepest lakes in the world are generally of tectonic origin, and their persistence through time has allowed the evolution of endemic plants and animals; that is, species that are found only at those sites.

Some of the best known examples of tectonic lakes and their endemic fauna are in the African Rift Valley, where each basin has been isolated for a sufficiently long time to allow adaptive radiation of numerous fish species. Lake Malawi has the largest number, with more than 850 fish. Most of these are endemic and are distributed among eleven families, with cichlids as the most species-rich. In Lake Tanganyika, sixteen families of fish are represented, including 200 cichlids, while the vast waters of Lake Victoria (68,800km^2; maximum depth 84m) are thought to have been the habitat for more than 500 fish species in the past. These lakes are facing pressure from agricultural development, fisheries, and new species introductions, for example of the Nile perch into Lake Victoria, and this combination of increased predation, competition, and changing water quality has resulted in the loss of many endemic species, perhaps as many as 200 in Lake Victoria.

Other ancient tectonic lakes with endemic species include Lake Biwa, Japan, with seventeen endemic fish, for example the giant Lake Biwa catfish *Silurus biwaensis*; Lake Titicaca, whose endemic fauna includes fifteen species of pupfish (*Orestias*) and a giant water frog (*Telmatobius culeus*); Lake Ohrid, with endemic sponges and fifty snail species; and Lake Baikal, which is the habitat for more than 1,000 endemic species including phytoplankton such as the diatom *Aulacoseira baicalensis*, invertebrates including amphipods and the dominant zooplankton species *Epischura baikalensis*, fish including the Baikal sculpin, and the sole exclusively freshwater species of seal, *Pusa sibirica*.

In terms of total numbers, most of the world's lakes (including those of the English Lake District; Figure 3) owe their origins to glaciers that during the last ice age gouged out basins in the rock and deepened river valleys. These lakes include the deep lakes of Europe such as Lake Geneva (310m deep), Lake Constance

3. Lakes of the English Lake District and their catchments. Situated in the northwest corner of England, the English Lake District (see Table 1) has been a major centre for freshwater studies from the 1920s onwards through the Freshwater Biological Association of the United Kingdom (FBA), and more recently through the Centre for Ecology & Hydrology (CEH). The lakes are arranged like the spokes of a wheel; this pattern is thought to be derived from a radial pattern of drainage over a central dome, which was then eroded and the valleys deepened by Pleistocene glaciers.

(251m), Lago Maggiore (372m), and Lake Como (425m); the freshwater lochs of Scotland including Loch Ness (227m) and Loch Morar (310m); the North American Great Lakes including Lake Michigan (281m) and Lake Superior (406m); and lakes in the South Island of New Zealand such as Lake Wakatipu (380m) and Lake Hauroko (462m). These glacial processes have also scratched out numerous shallower depressions in the ground,

Table 1 Lakes of the English Lake District (numbers relate to Figure 3)

Number	Lake	Area (km²)	Maximum depth (m)	Mean depth (m)
1	Windermere	14.8	64.0	21.3
2	Ullswater	8.9	62.5	25.3
3	Derwent Water	5.3	22.0	5.5
4	Bassenthwaite Lake	5.3	19.0	5.3
5	Coniston Water	4.9	56.0	24.1
6	Haweswater	3.9	57.0	23.4
7	Thirlmere	3.3	46.0	16.1
8	Ennerdale Water	3.0	42.0	17.8
9	Wastwater	2.9	76.0	39.7
10	Crummock Water	2.5	44.0	26.7
11	Esthwaite Water	1.0	15.5	6.4
12	Buttermere	0.9	28.6	16.6
13	Loweswater	0.6	16.0	8.4
14	Grasmere	0.6	21.6	7.7
15	Rydal Water	0.3	20.0	7.0
16	Blelham Tarn	0.1	14.5	6.8

for example on the Precambrian granites of northern Canada. Here the rocky landscape is replete with lakes that are just a few thousand years old, and the bottoms of many of these young Arctic waterbodies are coated with only the thinnest layer of lake sediments.

As the glaciers retreated, their terminal moraines (accumulations of gravel and sediments) created dams in the landscape, raising water levels or producing new lakes. These moraine-dammed, glaciated basins include the beautiful lakes of the lake district of southern Chile, such as Lake Llanquihue (317m; also affected by volcanic activity) and Riñihue Lake (323m). One of the largest (1,850km^2), deepest (586m), moraine-dammed lakes in South America is known by two names, since it extends across Patagonia from Argentina, where it is called Lake Buenos Aires, to Chile, where it is named General Carrera Lake. During glacial retreat in many areas of the world, large blocks of glacial ice broke off and were left behind in the moraines. These subsequently melted out to produce basins that filled with water, called 'kettle' or 'pothole' lakes. Such waterbodies are well known across the plains of North America and Eurasia.

As glaciers and expanding ice caps bulldoze their way across the landscape, their terminus may end in a basin of meltwater that is pushed ahead by the glacial ice flow. When the glaciers melt and retreat, these 'proglacial lakes' may expand dramatically until their ice dams dwindle in size or are breached. One of the most spectacular examples in North America is the twin set of proglacial lakes, Agassiz and Ojibway, that formed in front of the Laurentian Ice Sheet during the last Ice Age. At its maximum some 13,000 years ago, Lake Agassiz extended over an estimated 440,000km^2, almost twice the total area of the North American Great Lakes today. When a section of the ice cap finally collapsed in northern Hudson Bay around 8,200 years ago, the water gushed out and the merged Lake Agassiz-Ojibway drained almost completely, raising global sea levels by 0.8m or more. This catastrophic event is thought to have caused abrupt changes in oceanic circulation and climate, which in turn may have altered the human migration patterns and agriculture of prehistoric cultures in Europe.

The most violent of lake births are the result of volcanos. The craters left behind after a volcanic eruption can fill with water to form small, often circular-shaped and acidic lakes. The highest lake in the world is a small waterbody at 6,390m that occupies the crater of Nevado Ojos del Salado, an active volcano on the border of Chile and Argentina. Much larger lakes are formed by the collapse of a magma chamber after eruption to produce caldera lakes. The largest known of such lake-forming events was the supervolcanic eruption that formed Lake Taupo (at present 616km^2; depth 186m) in the central North Island of New Zealand, 26,500 years ago. This ejected more than 1,000 kilometres cubed (km^3) of material, and the resultant collapse created a huge caldera that filled with water. This lake has also undergone subsequent eruptions, most recently about 5,000 years ago, and today there are geothermal features within and near the lake attesting to the geologically active nature of the area.

Craters formed by meteorite impacts also provide basins for lakes, and have proved to be of great scientific as well as human interest. One of the most distinctive is Lake Manicouagan, a large (1,942km^2; depth 350m), ring-shaped lake in central Quebec caused by the impact of a 5km diameter asteroid some 214 million years ago. Much further north in the sub-Arctic region of Quebec, Canada, lies Pingualuk Lake, almost perfectly circular in shape with a diameter of 3km (Figure 4). This lake has long been known to the Inuit, who consider its crystalline waters to be imbued with healing powers and call it the 'Crystal Eye of Nunavik'. The crater it occupies is the result of a meteorite impact 1.4 million years ago, and because of its considerable depth (400m, with a current lake depth of 267m), the waterbody likely remained unfrozen beneath the thick ice sheet during glacial periods as a subglacial lake, perhaps with conditions similar to those found today in the subglacial lakes Vostok, Whillans, and Ellsworth in Antarctica. The deep sediments of Pingualuk Lake have been sampled by lake scientists to provide an uninterrupted record of past climates over several glacial–interglacial cycles.

There was a time when limnologists paid little attention to small lakes and ponds, but, this has changed with the realization that

4. Pingualuk Lake, the 'Crystal Eye' of northern Quebec.

although such waterbodies are modest in size, they are extremely abundant throughout the world and make up a large total surface area. Furthermore, these smaller waterbodies often have high rates of chemical activity such as greenhouse gas production and nutrient cycling, and they are major habitats for diverse plants and animals, including water fowl. A prominent example is Arctic thaw ponds and lakes ('thermokarst lakes') that are formed by the thawing of ice-rich, perennially frozen ground ('permafrost'). These occur in massive abundance over many parts of the northern landscape, with a total collective area of more than 250,000km^2. These waterbodies are undergoing rapid changes in response to permafrost thaw and degradation because of global climate change. In some areas they are disappearing as a result of evaporation, infilling, or drainage, while in other areas they are expanding in size and abundance. These waterbodies are hotspots of microbial activity that break down previously frozen stores of ancient soil carbon into carbon dioxide and methane, which are then released to the atmosphere.

The underwater shape of lakes

A bathymetric map showing the three-dimensional form or 'morphometry' of the basin is an essential starting point for any lake study. Such maps are increasingly available in digital format, yet there are still parts of the world where this basic information is completely lacking. Once a bathymetric map is obtained, several important values can be calculated. The first step is to calculate the area within each depth contour, which is most easily done with a geographic information system (GIS) software package. These areas can then be plotted on a graph as a function of depth. This area-depth plot is called a 'hypsographic curve', and it allows some useful statistics about the lake to be easily determined.

With the illustrative curve for Lake Baikal (Figure 5), we can ask the question: how much of this ancient lake has a depth greater than 500m? For most lakes of the world, the answer would be

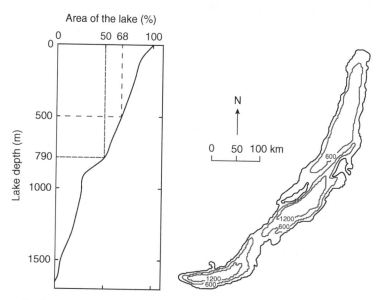

5. **Bathymetry and the area-depth curve for Lake Baikal, Russia.**

none, since even their maximum depth is much shallower. However, the Lake Baikal bathymetric map shows that it contains three deep basins, and the area-depth graph, which turns the complicated morphometry of the lake into a single curve, reveals that an impressive 68 per cent of the lake is deeper than 500m. Similarly, the curve can be easily read off to show that 50 per cent of the lake is 790m or deeper. The different depth layers in the hypsographic curve can be summed to calculate the total volume of the lake, which for Lake Baikal is 23,000km^3, equivalent to flooding all of England with freshwater to a depth of 176m. The total volume can then be divided by the area of the lake to give another limnological measure, the mean lake depth; for Lake Baikal this is 744m. In general, lakes with a larger mean depth tend to be of greater transparency and higher water quality, but human impacts on such lakes can cause a severe deterioration of these features, as has even been observed in the waters of Lake Baikal.

Rise and fall of water levels

In the simplest hydrological terms, lakes can be thought of as tanks of water in the landscape that are continuously topped up by their inflowing rivers, while spilling excess water via their outflow (Figure 6). Based on this model, we can pose the interesting question: how long does the average water molecule stay in the lake before leaving at the outflow? This value is referred to as the water residence time, and it can be simply calculated as the total volume of the lake divided by the water discharge at the outlet. This lake parameter is also referred to as the 'flushing time' (or 'flushing rate', if expressed as a proportion of the lake volume discharged per unit of time) because it provides an estimate of how fast mineral salts and pollutants can be flushed out of the lake basin. In general, lakes with a short flushing time are more resilient to the impacts of human activities in their catchments, although they are certainly not immune to such effects.

$$\text{Residence time} = \frac{\text{Lake volume}}{\text{Outflow}}$$

Lake	Residence time
Lake St-Charles	30–100 days
Windermere	9 months
Lake Geneva	11 years
Lake Baikal	330 years
Lake Titicaca	1200 years

6. Differences in water residence time among lakes.

Each lake has its own particular combination of catchment size, volume, and climate, and this translates into a water residence time that varies enormously among lakes (Figure 6). Lake St-Charles, for example, is our drinking water reservoir in Quebec City, and is a dammed river basin that is fed by a large catchment (169km^2) relative to the size of the lake (3.6km^2). The residence time for this reservoir is therefore only one to a few months, depending on the time of year. At the other extreme, Lake Titicaca has a small discharge relative to its volume, and its calculated residence time is more than 1,000 years.

A more accurate approach towards calculating the water residence time is to consider the question: if the lake were to be pumped dry, how long would it take to fill it up again? For most lakes, this will give a similar value to the outflow calculation, but for lakes where evaporation is a major part of the water balance, the residence time will be much shorter. This is the case for Lake Titicaca, where the true water residence time is only eighty years based on inputs (not the 1,200 years based on outflow, as in Figure 6), because more of its water is lost by evaporation than through the outflow. This lack of complete flushing also means that salts are concentrated in the lake, and the water is slightly brackish (salinity around 0.7 parts per thousand (ppt)).

If the total volume of water entering the tank in Figure 6 is the same as the volume leaving via its outflow spigot, then the water level will remain constant. For lakes, however, this is often not the case, and residents near a lake shore will be aware of the rapid, sometimes alarming, fluctuations in water level that can occur after a heavy rainfall or other events. One extreme example is Riñihue Lake in southern Chile, where a powerful earthquake in 1960 triggered landslides that blocked the outflow. The water level rose by 20m. With the threat of serious flooding damage if the waters breached the dam, plans were made to evacuate 100,000 people in the downstream city of Valdevia and surroundings. Fortunately, much of this water could be released in a controlled fashion by excavation of release channels over the subsequent weeks.

Large variations can also occur as a result of the natural cycles of river flow. The most pronounced example is the Amazon River and its extensive floodplain, known as the 'varzéa'. Many fish depend upon the annual flooding cycle, which allows them to swim into the inundated forest to feed on terrestrial insects, spiders, nuts, seeds, and flowers. One of the numerous lakes is Lago Calado, which lies 60km upstream of Manaus, the Brazilian city at the confluence of the Rio Solimões and the Rio Negro, in the heart of the Amazonian rainforest. Like other lakes of the region, it has a complex, dendritic shape, and when its basin floods with the cappuccino-coloured waters of the Rio Solimões, the lake level rises by 10m, and its area expands by a factor of four. Varzéa lakes contain floating meadows of the grasses *Paspalum repens* and *Echinochloa polystachya*, and these rise and fall with the seasons, providing lush island habitats for insects, birds, and snakes.

Climate change has a major effect on lake water levels by altering the balance of inflows and evaporative losses. One of the most striking examples is Lake Chad, at the edge of the Saharan Desert. Given its shallow depth (maximum depth 11m; mean depth 1.5m), it is highly sensitive to variations in rainfall, both with season and in the longer term. Over the last fifty years, there has been a

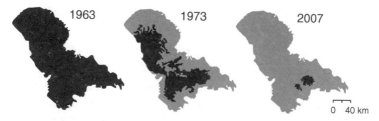

7. **The shrinking of Lake Chad in central Africa caused by ongoing dry conditions.**

massive contraction in lake area (Figure 7) associated with a drier climate, compounded by inefficient damming, irrigation, and land clearance for agriculture. Violent conflicts have emerged between fishermen and farmers, who have opposing needs for the water. The geological record shows that Lake Chad has undergone profound changes in the past, from palaeolake Mega-Chad of more than 1 million km^2, to periods of almost complete dryness. The full loss of this resource today would have devastating consequences for more than thirty million people who currently depend on water from the lake.

Drops in lake level can also reveal some surprises. Lake Kinneret (area $167km^2$; maximum depth 43m) in Israel is also known as the Sea of Galilee, a place that figures prominently in the New Testament of the Bible. During a period of drought in the late 1980s, the water level of this freshwater lake dropped 9m and revealed the presence of a stone-age settlement of oval huts (an archaeological site now called Ohalo II), dating back 23,000 years before the present. These are among the oldest known dwellings in the world, and are evidence of humankind's longstanding association with lakes.

Lake sediments as archives

Lake Chad is an extreme example of how lakes respond to environmental change, but even subtle variations in climate and human activities are registered in lakes, and in a way that can be decoded by careful analysis. Each year, mineral and organic particles are deposited by wind on the lake surface and are washed in from the catchment, while organic matter is produced within the lake by aquatic plants and plankton. There is a continuous rain of this material downwards, ultimately accumulating as an annual layer of sediment on the lake floor. These lake sediments are storehouses of information about past changes in the surrounding catchment, and they provide a long-term memory of how the limnology of a lake has responded to those changes. The analysis of these natural archives is called 'palaeolimnology' (or 'palaeoceanography' for marine studies), and this branch of the aquatic sciences has yielded enormous insights into how lakes change through time, including the onset, effects, and abatement of pollution; changes in vegetation both within and outside the lake; and alterations in regional and global climate.

The sampling approach in palaeolimnology begins with the coring of the lake sediments. A variety of portable devices have been developed to obtain short cores for analysis of the last few hundred years of the sediment record, while much heavier drilling equipment is used to core the sediment to much greater depth, and to extend the record back to tens of thousands of years or longer. For example, at Lake El'gygytgyn, a crater lake in Siberia formed by a meteorite impact 3.58 million years ago, a 400m long sediment and rock sequence was obtained to capture a continuous 3.6 million year record of Arctic climate change, including the Pliocene–Pleistocene transition. At Lake Biwa, Japan, formed by tectonic processes three million years ago, the upper 250m of lake sediments from a 1.4km long core provided records that extend back to 430,000 years before the present.

Sampling for palaeolimnological analysis is typically undertaken in the deepest waters to provide a more integrated and complete picture of the lake basin history. This is also usually the part of the lake where sediment accumulation had been greatest, and where the disrupting activities of bottom-dwelling animals ('bioturbation' of the sediments) may be reduced or absent. Once the core is brought up to the surface, the sediment is extruded through its core barrel and split into sections. The age of the upper layers is established by measuring the radioisotopes cesium-137 and lead-210, which provide a chronology of sediment dates for the last 150 years or so. For deeper, older layers, the dates are established by analysis of the radioisotope carbon-14. With several of these radioisotopic analyses down the core, dates can then be estimated by interpolation for each and every layer, and these time-stamped strata then analysed for evidence of change.

At first inspection under the microscope, a sample from one of those core slices will seem little more than a smear of nondescript particles that vary randomly in shape and size. However, with careful observation, the undecomposed remains of all manner of terrestrial and aquatic life can be discerned among the particles, and in many cases identified to their species of origin. These 'microfossils' include pollen grains that have been blown or washed into the lake, and that have resisted decomposition in the sediments because of their hardy outer walls. The distinctive shape of the pollen grains allows them to be identified to genus and even species, and the lake sediments thereby provide a record of changes in plant community structure in the surrounding landscape.

Some of the most informative microfossils to be found in lake sediments are diatoms, an algal group that has cell walls ('frustules') made of silica glass that resist decomposition. Each lake typically contains dozens to hundreds of different diatom species, each with its own characteristic set of environmental preferences as well as distinctive cell wall shape and ornamentation that can be used to identify them from their microfossil remains in the sediments. A widely adopted approach is to sample many lakes and establish a statistical relationship or 'transfer function' between diatom species composition (often by analysis of surface sediments) and a lake water variable such as temperature, pH, phosphorus, or dissolved organic carbon. This quantitative species–environment relationship can then be applied to the fossilized diatom species assemblage in each stratum of a sediment core from a lake in the same region, and in this way the physical and chemical fluctuations that the lake has experienced in the past can be reconstructed or 'hindcast' year-by-year. Other fossil indicators of past environmental change include algal pigments, DNA of algae and bacteria including toxic bloom species, and the remains of aquatic animals such as ostracods, cladocerans, and larval insects.

One example of tapping into these historical archives is a study at Walden Pond (Figure 8), a kettle lake near the city of Boston in Massachusetts, USA, with a maximum depth of about 30m and surface area of 25 hectares (ha). This is a site well known to readers of American literature, because it was here that naturalist, essayist, philosopher, and historian Henry David Thoreau spent two years on its shores, from 4 July 1845 to 6 September 1847. The experience formed the basis of his classic work *Walden*, published in 1854, and it inspired his meditation on the natural world, writing how 'A lake is a landscape's most beautiful and expressive feature. It is earth's eye; looking into which the beholder measures the depth of his own nature'.

Thoreau kept detailed records in his daily journal of many features of the lake, including notes on the layering of the water,

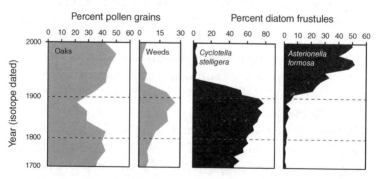

8. **Environmental changes over the last 300 years recorded in the lake sediments of Walden Pond, USA.**

with warm above and cold below. This was some twenty years before Forel began his studies at Lake Geneva, and half a century before James G. Needham, Professor of Limnology at Cornell University, wrote the first English textbook on freshwater ecology. It has been suggested that Thoreau, despite his reluctance to embrace all aspects of science, might well be considered North America's first limnologist.

The pollen profiles in the 28-centimetres (cm) long sediment core shown in Figure 8 bear witness to the clearing of forest by the early settlers of New England, with a reduction in oaks and the expansion of agricultural lands, hence an increase in weed pollen. It is ironic that while Thoreau was making his observations about the spiritual value of lakes and forests, this wholesale destruction of forests was at its peak, with 80 per cent of the land ultimately clear-felled and converted to agriculture. Towards the beginning of the 20th century, a reversal of this process is apparent in the pollen records, as rural populations left their homes to work in cities and allowed the uneconomic farmlands to revert back to forest, as indicated in the sediment records by a reduction in weed pollen and a proportionate increase in oak pollen.

The diatom analyses show that other changes were in motion at Walden Pond in the period 1880 to the present (Figure 8). Certain species such as *Cyclotella stelligera* rose gradually in prominence, and then suddenly gave way to species such as *Asterionella formosa* that are more characteristic of nutrient-enriched waters. The application of a transfer function for phosphorus to the full assemblage of diatoms in the sediments indicated that this key nutrient stimulating algal blooms rose abruptly in concentration from the 1920s onwards, associated with the recreational development of the lake. Thoreau's celebration of nature while living alone at Walden Pond is now shared by thousands of visitors each summer, and ongoing careful management will be required to preserve the cultural and ecological values of this iconic lake for future generations.

Chapter 3
Sunlight and motion

> When the waves are lively...the proportion of blue mingles with the reflected colours, and varies with the shape and size and direction of the waves.
>
> F. A. Forel

François Forel devoted much of his treatise on Lake Geneva to the physical environment of the lake: light, temperature, wind, waves, currents, and mixing. He noticed that water at times moved in and out through a narrow opening to the harbour near his house at Morges with surprising speed and regularity, and he realized that this had something to do with a see-saw rocking motion that extended across the entire lake. He talked with fishermen, who told him strange stories about how their nets set deeply beneath the surface were dragged by currents, but in the opposite direction to that of the wind. He realized that the lake was not simply a basin of well-mixed water, like an aerated fish tank, but instead was composed of layers with different temperatures, and that this layering changed according to season.

Forel was especially fascinated by the interplay between sunlight and water, an interest encouraged by watching his friend the painter François Bocion at work as he captured Lake Geneva's multi-coloured hues of sky, clouds, and water on the canvas. Forel noted changes in the turbidity of inshore waters, from clear to opalescent to muddy, and he surmised that water transparency could be a simple yet powerful indicator of the state of health of a lake ecosystem. Freshwater scientists today are very much aware of the importance of all of these features, which define the physical habitat characteristics of lakes and strongly influence their chemistry, biology, and ecosystem services (Figure 2).

Clear or murky waters

Soon after Forel began his studies at Lake Geneva, he learned of a simple method to measure water transparency, and he became the first person to apply it to lakes and develop a standardized protocol. The method had been conceived for the clear blue waters of the Mediterranean Sea by a priest and scientific advisor to the Pope, Pietro Angelo Secchi. Working on board the Papal Navy ship 'Immacolata Concezione', Secchi undertook studies to better understand the interaction between sunlight and the sea. His elegant approach was to simply lower a white disc and note the depth at which it could no longer be seen.

Forel seized upon this idea from Secchi and he established a formal protocol: take a 20cm diameter white disk, note the depth where the Secchi disk disappears, bring it up again slowly noting the depth at which it reappears, and calculate the 'Secchi depth' as the average of those two values. Secchi had used disks of different colours and sizes, including one that was 2.37m in diameter. Forel recommended 20cm because of its portability for travelling, and also because he saw little difference with a larger, 35cm version. He employed both a white painted, zinc metal disk and a white glazed, ceramic dinner plate, and noted that while the former was less fragile, the latter retained its white coloration for longer. These days, 20cm or 30cm diameter metal disks are routinely used in lake studies, and are usually painted with alternating quadrants of black and white to enhance contrast.

Values of Secchi depth range from a few centimetres in highly polluted waters with algal blooms, to tens of metres in the world's most transparent lakes. The overall record is held by the Weddell Sea, Antarctica, where a 20cm diameter Secchi disk could be seen to a depth of 79m; this is close to the theoretical limit for visibility in pure water. For lake waters, the record is Crater Lake, Oregon, USA, where a 1m diameter Secchi disk (somewhat outside Forel's specifications, but fine for Father Secchi) could be seen to 44m.

In lake and ocean studies, the penetration of sunlight into the water can be more precisely measured with an underwater light meter (submersible radiometer), and such measurements always show that the decline with depth follows a sharp curve rather than a straight line (Figure 9). This is because the fate of sunlight streaming downwards in water is dictated by the probability of the photons being absorbed or deflected out of the light path; for example, a 50 per cent probability of photons being lost from the light beam by these processes per metre depth in a lake would result in sunlight values dropping from 100 per cent at the surface to 50 per cent at 1m, 25 per cent at 2m, 12.5 per cent at 3m, and so on. The resulting exponential curve means that for all but the clearest of lakes, there is only enough solar energy for plants, including photosynthetic cells in the plankton (phytoplankton), in the upper part of the water column.

The depth limit for underwater photosynthesis or primary production is known as the 'compensation depth'. This is the depth at which carbon fixed by photosynthesis exactly balances the carbon lost by cellular respiration, so the overall production of new biomass (net primary production) is zero. This depth often corresponds to an underwater light level of 1 per cent of the sunlight just beneath the water surface (Figure 9). The production of biomass by photosynthesis takes place at all depths above this level, and this zone is referred to as the 'photic zone'. At deeper,

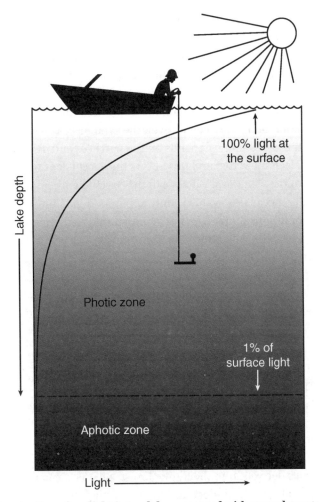

9. **Penetration of sunlight into a lake measured with an underwater light meter.**

less illuminated depths, no photosynthetic growth is possible and biological processes in this 'aphotic zone' are mostly limited to feeding and decomposition.

A Secchi disk measurement can be used as a rough guide to the extent of the photic zone: in general, the 1 per cent light level is about twice the Secchi depth. However, this is not always accurate because the stream of photons passing down through the water, hitting the Secchi disc, and returning to our eyes at the surface is affected by the two different processes: absorption (given the symbol 'a') and deflection (called 'scattering' and given the symbol 'b'). In combination, these processes dictate the overall attenuation of light, which is given the symbol 'c'. The relative importance of a and b depends on the particles and dissolved materials in the water, and this affects the Secchi depth. Specialists in the optics of natural waters call the Secchi depth an 'apparent property', because its value is only apparent depending on the light conditions under which it is measured; the Secchi will be shallower in the late afternoon, for example, because of the lower sun angle, and of course it will be barely seen at all at night, even under the brightest moonlight. On the other hand, a, b, and c are called 'inherent properties' because they are intrinsic characteristics of the lake water that are not affected by sunlight conditions at the time of measurement.

In general, more algal particles in the water cause more absorption and scattering, and the Secchi depth is a measure of nutrient pollution and algal enrichment of a lake ('eutrophication', described in Chapter 7). However, for lakes in heavily forested regions such as the circumpolar boreal zone and the Amazonian basin, the inflowing waters are highly charged in brown, tea-like materials derived from the humic forest soils. These materials strongly absorb the underwater light, masking the effects of algae. At the opposite extreme are waters with lots of reflective mineral particles in suspension. These waters have high b values, and many of the photons reflected by the Secchi disc are scattered out of the light beam that comes back to our eyes, but they are still available for photosynthesis in the lake. This means that the Secchi depth has to be multiplied by a larger number (up to three in some waters laden with mineral particles) to estimate the extent of the photic zone.

Despite its limitations, the Secchi disk has proved to be an extremely valuable tool for lake studies and science communication. Charles R. Goldman, Professor of Limnology at the University of California Davis, initiated a long term study of Lake Tahoe in the 1960s based on a broad spectrum of measurements including nutrients, oxygen, plankton biomass, and photosynthesis. He noted that of all these limnological records, it was the reduction in Secchi depth through time that provided the most easily understood and convincing evidence for policy makers that stringent controls on watershed management were needed to protect the renowned clarity and blueness of the lake. The Secchi disk continues to be used routinely in lake studies (generally in combination with submersible optical instruments), and given its low cost and simplicity, it is also used in many parts of the world as part of public outreach activities and citizen monitoring programmes. The North American Lake Management Society (NALMS), for example, runs 'The Secchi Dip-In' every year that involves hundreds of lake shore residents and visitors throughout the USA and Canada.

Water colours

Forel took a special interest in how the colour of water differed among lakes and even among various parts of the same lake. He developed a liquid colour scale to classify different waters (today available as a smart phone app), and he undertook experiments to try to understand the reasons for these differences. Little could he imagine that water colour is now used in so many powerful ways to track changes in water quality and other properties of lakes, rivers, estuaries, and the ocean. A broad suite of optical technologies is increasingly available, ranging from profilers that can be lowered down into the lake to measure different spectral bands (including UV-radiation), to moored systems that automatically take measurements throughout the year, and satellites that continuously monitor changes in lake water colour from space.

Lakes have different colours, hues, and brightness levels as a result of the materials that are dissolved and suspended within them. The purest of lakes are deep blue because the water molecules themselves absorb light in the green and, to a greater extent, red end of the spectrum; they scatter the remaining blue photons in all directions, mostly downwards but also back towards our eyes. Some lake waters such as Lake Vanda in Antarctica and Crater Lake in Oregon are so deeply inky-blue that it almost seems as if dipping your hand in them would stain it indigo.

Algae in the water typically cause it to be green and turbid because their suspended cells and colonies contain chlorophyll and other light-capturing molecules that absorb strongly in the blue and red wavebands, but not green. However there are some notable exceptions. Noxious algal blooms dominated by cyanobacteria are blue-green (cyan) in colour caused by their blue-coloured protein phycocyanin, in addition to chlorophyll. Some village ponds in England have been called 'traffic light ponds' because they change from green to red over the course of the day. This is because their algae (classified in a group of motile species called Euglenophyta) contain red-coloured particles in addition to their photosynthetic pigments. In dim light and darkness, these red particles are hidden within the cellular interior, and the bright green chloroplasts are fully exposed towards the outside environment. When the sun is shining brightly, however, the red particles migrate outwards to mask the colour of the chloroplasts and protect the cells against solar damage. Some aquaculture fish pond managers have been startled to see their euglenophyte-containing ponds suddenly turn bright red in the sun. Red-coloured water may also be produced by other freshwater algae, for example *Uroglena* that forms red tides, cyanobacteria of the genus *Planktothrix* containing the red protein phycoerythrin, and the blood-red coloured alga *Haematococcus* that is common in garden bird baths throughout the world.

The yellowness that Forel observed in the inflows to Lake Geneva and its nearshore waters is due to the dissolved organic materials derived from the catchment. Specifically it is caused by the complex mixture of high molecular weight chemicals called 'humic acids': tea-like substances that are derived from the breakdown of leaf litter in the soils and that are then washed into the lake. These materials used to be called 'gelbstoff', which literally translates from the German into English as the imprecise term 'yellow stuff'. Forel was dead right when he said in 1895: 'What is the nature of these organic materials in the lake water? This question has not been sufficiently studied'. More than a hundred years later, this is still an active area of study in lake and ocean science. These days, the golden material is called 'coloured dissolved organic matter' (CDOM, pronounced 'see-dom'). This modern term again has a certain vagueness about it, concealing our still limited knowledge about the exact chemistry of this complex mixture.

One of the many interesting features of CDOM is that it strongly absorbs blue light, and absorbs even more strongly in the ultraviolet radiation (UV) part of the sunlight spectrum. It is therefore the natural sunscreen of lakes and rivers, protecting aquatic biota from harmful UV burn. The effect of CDOM on lake colour is all a matter of concentration. At highest concentrations, it absorbs sunlight at all wavelengths and the water is stained black, like espresso coffee. At lower, more typical, concentrations, the CDOM absorbs blue and blue-green light, leaving the water a golden brown colour, while at the lowest concentrations, CDOM absorbs the blue wavelengths, the water molecules themselves absorb the yellow to red wavelengths, and all that is left is the green waveband of light that is then perceived by our eyes. Forel proved this to himself by filtering brown, CDOM-rich swamp water, diluting it in clear lake water from Lake Geneva, and then filling a long, glass-bottomed tube that he could look through to the sky above; the water was transparent lime-green in colour, just as he had observed in the inshore region of Lake Geneva where inflowing streams mix into the lake.

Mysterious water

Water is such a common part our daily lives that we take it for granted, and certainly do not think of it as a chemical each time we turn on the tap or take a sip of something liquid. Yet it is a chemical substance with strange properties, some of which still elude a full explanation. These properties have enormous consequences for the physical and chemical nature of lakes, and for the aquatic life that lives within them.

At the heart of this strangeness is the water (H_2O) molecule itself and its tendency to aggregate in ever-changing clusters of different sizes and complexity. An electron is shared between each hydrogen atom and the oxygen atom, thereby forming a covalent bond that keeps the molecule together. But with eight times the positively charged protons than hydrogen, which has only one proton, the oxygen atom is the big brother in this relationship, and draws the shared, negative electron cloud slightly away from the hydrogen. The result is that the oxygen has a slight negative charge, while the two hydrogen atoms are left with slight positive charges. Opposites attract, so the molecules of water stick together, with the electrostatic O^-—H^+ interactions among them called hydrogen bonds (Figure 10). Each water molecule can hydrogen bond to a maximum of four others, and although the debate is ongoing, most H_2O molecules in liquid water appear to be dynamically linked together in pyramid-like clusters (tetrahedrons), with one of the oxygen atoms at the centre of the pyramid.

Another weird attribute of H_2O is its peculiar density versus temperature relationship. Solids tend to be more compact and dense relative to their liquid form, yet ice is exactly the opposite and floats on liquid water. This is because all of the H_2O

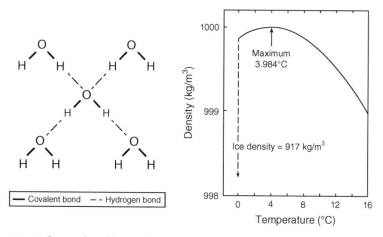

10. Hydrogen bonding and the strange density–temperature relationship for water.

molecules in ice are hydrogen bonded to four others with maximal distance among molecules in this crystal array. Once the ice melts, the liquid H_2O loses this expanded structure and full hydrogen bonding of all molecules; the molecules become more closely packed, resulting in greater density. This relaxing of structure continues with increased warming until around 4°C (3.984°C to be precise, at atmospheric pressure), which is the temperature of maximum density for pure water. With further warming above that, the increasing kinetic energy and motion of the water molecules result in an increasing average distance between them (although never to the extent of that in ice), and the density decreases (Figure 10).

Why is this density–temperature relationship so important? For Canadians and other northerners, it means that we can travel by ski, snow-shoe, and snowmobile over lakes in winter, confident in the knowledge that an impressively solid form of H_2O floats at the surface and will (hopefully) support us above the frigid liquid beneath. Of broader significance throughout the world, it also means that warm water floats on cold water, so that when the lake heats up in summer, a layer of warm water will float over the top of cold dense water at the bottom. The sharp temperature change between these two layers is called the 'thermocline', and the upper and lower layers that it separates are called the 'epilimnion' and 'hypolimnion', respectively. These two strata differ not only in temperature, but also in their chemistry and biology.

Lake seasons and mixing

From ice-covered waters to warm water floating on cold, the extent of layering of lakes, called 'stratification', varies greatly with season. At any one time, the layers may have strikingly different physical and chemical properties. For example, at Lake St-Charles, our drinking water reservoir in Quebec City, Canada, the thermocline in late summer lies at 7–10m depth (Figure 11). There is not only a sharp drop in temperature over this depth range, but also an abrupt decline in oxygen. The epilimnion is well charged in oxygen by exchange with the overlying atmosphere, but the thermocline acts as a barrier to that exchange, and the

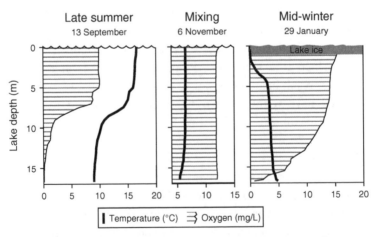

11. **Seasonal changes in Lake St-Charles, the water reservoir for Quebec City. The bottom scale is for both temperature (thick lines, values in degrees Celsius (°C)) and oxygen (shaded areas, values in milligrams per litre (mg/L)).**

bottom waters are completely depleted in this life-supporting gas. The hypolimnion of Lake St-Charles in late summer to early autumn is therefore a singularly unattractive place for fish such as trout, which only thrive in highly oxygenated waters.

With autumnal cooling of the epilimnion, the temperature difference between the top and bottom of the lake becomes much less, and there is no longer the strong density barrier against mixing between the surface and bottom layers. Additionally and most importantly, as the lake surface cools, the water becomes colder, denser, and sinks, giving rise to convective circulation that acts in concert with the wind-induced mixing. Eventually the entire water column mixes, called lake overturn, and the bottom waters are recharged with oxygen from the atmosphere (Figure 11). The solubility of all gases, including oxygen, is greater in cold water, and so by the end of this mixing period, the O_2 concentration increases to well above that in the warm summer epilimnion.

Temperate lakes in milder climates may be too warm to freeze in winter. These lakes are called 'monomictic' because they have only one, albeit long, period of vertical mixing, and atmospheric oxygen continues to recharge the lake as the water cools and mixes from autumn through winter. Many lakes of the world are monomictic, including lakes Biwa, Geneva, Tahoe, Titicaca, Taupo, and Maggiore, and lakes of the English Lake District.

The solubility of oxygen in water, even cold water, is not that great relative to the demand for it by chemical and biological processes, and in all lakes there is a precarious balance between income and expenditure columns of the ledger for this vital gas. This is nowhere more apparent than in northern, temperate lakes that have an ice cap in winter, such as Lake St-Charles (Figure 11). Although such lakes are well charged in oxygen due to pre-winter cooling and mixing, the winter ice and its associated snow cover cuts off light for oxygen generation by photosynthesis and eliminates the possibility of oxygenation from the overlying atmosphere. Decomposition meanwhile continues to consume oxygen, especially in the sediments, and can eventually drive the bottom waters to complete absence of oxygen (i.e. anoxia). Throughout winter, the lake is highly layered, this time with the colder (but less dense) water floating over the somewhat warmer water beneath. This is referred to as 'inverse stratification' because of the inverted temperature profile (cold over warmer), but the water density increases with depth, as it must to be in gravitational equilibrium.

In spring, the ice melts, Canadians pack away their snow boots, and the frigid surface layer of the lakes warms to the temperature of the waters beneath. With little difference in temperature and therefore density between the surface and deeper waters, the wind is able to mix and re-oxygenate the entire lake from top to bottom. Such lakes therefore have a second season of mixing, and are referred to as 'dimictic lakes', with two full water column mixing periods each year: autumn and spring. Unlike in autumn, however, spring mixing is rapidly dampened by the ongoing seasonal changes in temperature; once the surface water warms above 4°C it becomes less dense than the cold spring condition, and begins to float at the surface as an ever warming layer that impedes further mixing. For this reason, the spring mixing period in dimictic lakes can be brief and sometimes barely apparent relative to the prolonged period of mixing in autumn.

Lively waves at the surface

When the wind gently blows across a lake, it roughens the water surface to produce ripples. These wavelets are dragged upwards by the friction of the wind, and they subside into troughs as a result of the hydrogen bonding that pulls water molecules back down into the lake. They are referred to as 'capillary waves', to formally acknowledge that their restoring force is the capillary molecular interaction (i.e. surface tension) of the water. They have a maximum wavelength of 1.73cm and a period of less than a second. As the wind builds, the waves are dragged higher and now the restoring force is dominated by gravity. At winds above 25–30km per hour, or as the waves move into shallower depths inshore, the tops of the waves start travelling faster, overextend the base of the waves and break to produce whitecaps, causing intense mixing and oxygenation of the surface waters. Gravity waves have been observed up to 8m amplitude (i.e. trough to crest) in the North American Great Lakes, but lakes do not have the vast wind fetch of the ocean and most lake waves are less than 50cm in height.

At first impression, it seems that surface gravity waves have a large amount of energy that should mix the lake. The waves do give rise to movement below the surface, specifically a series of circular motions that decrease in diameter exponentially with depth, but they are not enough to fully mix the water column. This wave action causes the re-suspension of fine sediments in the littoral (i.e. inshore) region of the lake, which means that fine sediments only accumulate offshore, in deeper water, where the wave effects cannot penetrate. The depth threshold between re-suspension and sediment deposition is called the 'mud depositional boundary depth', and it depends on the wave height generated by storm events and on the slope of the inshore zone. Greater mixing of the lake, however, depends not on these 'lively waves', but rather on slower waves that are much less obvious to most lake visitors.

Slow waves at and below the surface

Forel noted in his autobiography that one of his favourite subjects of research was studying the slow, pendulum-like rocking motion of lakes. This phenomenon is so well known at Lake Geneva that the residents gave a name to it in the local Swiss-French dialect; for many centuries they have called it a 'seiche' (pronounced 'say-sh'), a term that is now used by scientists around the world to describe this ubiquitous feature of lakes. The most obvious aspect of seiches is the change in lake level, which rises and falls over a period of minutes to hours, particularly at the shore.

Forel made detailed observations of seiches at Lake Geneva and elsewhere with various types of continuous lake level recorders that he had conceived and installed, including a portable version. He was frustrated with his initial attempts to derive a mathematical theory of seiches, and noted with regret in his autobiography that while at college he had abandoned a useful class in differential calculus, as the teacher was particularly uninspiring (this itself is a lesson to all professors).

Forel did, however, have many other talents, including the enthusiasm and ability to network with other scientists throughout Europe and the world. He made contact with one of the leading physicists of the day, William Thomson, later named Lord Kelvin, who helped him simplify an initial unwieldy equation into the elegant form:

$$P = 2L / \sqrt{(gh)}$$

where P is the period of the rise and fall in lake level, L is the length of the lake, g is the gravitational constant, and h is the average depth of the lake. Forel correctly surmised that the seiche is a standing wave that extends across the full length of the lake, and that it could also be accompanied by lower amplitude, secondary waves.

So what is the origin of these oscillations? Forel rightly concluded that seiches begin by a consistent wind that blows and pushes the water towards one end of the lake (Figure 12). This 'set-up' condition results in the lake level being higher downwind and lower upwind, but it is not stable, and as soon as the wind stops, the water rocks back and overshoots in the opposite direction. This see-saw motion will continue until all the potential energy

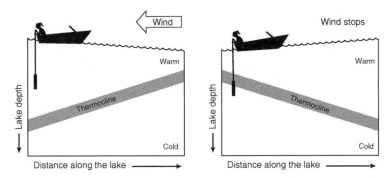

12. The surface seiche is caused by the wind pushing water to one end of the lake, and the resultant internal seiche can be detected as an oscillation of the thermocline.

associated with the initial displacement of the lake surface is finally dissipated, just as a pendulum finally comes to rest.

Forel would have been amazed to learn that the greatest significance of seiches lies well beneath the lake surface, at the depth of the thermocline where waves and mixing affect the transfer of oxygen and nutrients. He had evidence from his own work that the thermocline could change in depth over short periods of time, but he did not at the time make the link with the surface variations in water level. It was not until the classic work by Ernest M. Wedderburn and others in the lochs of Scotland that the nature of this 'internal wave' or 'internal seiche' came to be revealed. When the water piles up at the downwind end of a lake, the greater overlying mass of water in the epilimnion pushes the thermocline downwards. When the wind stress is relaxed, the thermocline rises up again, overshoots, and continues to oscillate until the set-up energy is dissipated.

Figure 12 captures the general idea of this wind-induced motion of the thermocline, but it needs some qualification. The vertical scale has been exaggerated and it does not express an important feature: that the internal seiche is slower and has a period that is much longer than that of the surface seiche. In Lake Geneva for example, the period for the surface seiche along the main axis of the lake is about seventy-four minutes, while the main internal wave (which again can be accompanied by additional modes of smaller waves of higher harmonics) has a period of three days, and it keeps going long after the initial setup by the wind and the dissipation of the surface seiche. The freshwater scientist shown in the figure would need to remain anchored on station for many hours to days to properly observe this slow moving internal wave. She would then notice from her underwater thermal profiles that the temperature at any specific depth gradually oscillates up and down, with the greatest variations in and near the thermocline.

Internal waves continue to be of intense interest to freshwater researchers, for many reasons. First, there is the question of size: they can be huge. Surface seiches are displacements of the air–water interface, and given the large difference in density of these two fluids, it takes a lot of wind energy to cause even a small increase in water level (thus potential energy) at the downwind end; for this reason, the amplitude of surface seiches is generally small, of the order of cm to tens of cm, although exceptions are known; for example, 5m seiches associated with storm surges on Lake Erie. For the internal wave, on the other hand, the density differences between the epilimnion and hypolimnion are small, and the same potential energy associated with a given surface displacement translates into a massive displacement at the level of the thermocline. During the set-up phase of internal waves in deep lakes such as Lake Tahoe, the bottom water can be displaced vertically by as much as 100m, as an 'upwelling' event near the upwind shore; this brings nutrients up into the photic zone, where they can stimulate algal production.

The motion of water in lakes and seas is affected by the rotation of the Earth, and internal waves are no exception: as the waves oscillate on the thermocline, they are deflected to the left in lakes of the Southern Hemisphere and to the right in lakes of the North Hemisphere. This so-called 'Coriolis Effect' is relatively weak, but in medium to large lakes it can be observed to modify the internal waves in two ways. First, it causes waves to be trapped and guided around the edge of the lake, in anticlockwise (Northern Hemisphere) or clockwise (Southern Hemisphere) direction. Lord Kelvin was the first to discover this phenomenon that is found in the atmosphere and ocean as well as in large lakes, and to define it formally. It is also fitting that the term 'Kelvin wave' makes reference to the physicist who helped Forel develop his theory of seiches (and perhaps Lord Kelvin was stimulated in his own thinking about waves during these discussions), for in many lakes this is a feature with major consequences. In Lake Biwa, Japan, for example, the Kelvin wave can be so highly energized by the strong winds of the typhoon season that it sometimes attains the surface, bringing cold, nutrient-rich waters into the upper layers as it circulates around the lake.

The second type of internal wave affected by the Coriolis force occurs offshore, in the main body of the lake, and is called a 'Poincaré wave', named after the brilliant French mathematician and theoretical physicist Henri Poincaré. These waves can similarly be of high amplitude. The example shown in Figure 13 for Lake Ontario, Canada–USA, is from temperature measurements at a fixed sampling station over a five-day period. The wave-like nature of the observations is striking, as is the amplitude of oscillation,

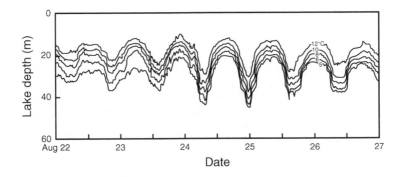

13. **Poincaré waves on the thermocline of Lake Ontario.**

which reached 25m from trough to crest. Poincaré waves have a much shorter period than the inshore Kelvin waves, but still much longer than the surface seiche; in Lake Ontario this period is sixteen hours (Figure 13) versus ten days for the Kelvin wave and five hours for the surface seiche along the main axis of the lake.

Lake biologists and biogeochemists all have a special interest in internal waves because these motions play a decisive role in mixing the water between layers, which, as seen in Figure 11, can differ greatly in temperature and oxygen, as well as in many other properties. The wave-induced, horizontal movement of water extends to the lake floor: as the water mass slides backwards and forwards within the lake basin, an oscillating turbulent current moves across the sediments and can bring particles up into suspension into a bottom zone called the 'benthic boundary layer'. This water flow and mixing drives oxygen into the sediments, and it accelerates exchanges of other chemicals such as nitrogen and phosphorus between the lake bottom and the overlying water. Of special relevance to the plankton, the tilting of the thermocline during set up and subsequent oscillations brings deep water closer to the surface, and it exposes that water to surface mixing. This entrains nutrients from depth into the photic zone, and gives rise to horizontal flows that help mix and homogenize the lake.

And then there are 'billows' (Figure 14). The internal seiche generates water movements that are in opposing directions above and below the thermocline. This creates friction and a pressure

14. **Billows across the thermocline. These begin as progressive waves that roll up, break, and mix materials between the epilimnion and hypolimnion.**

gradient that results in short-period, progressive waves that propagate along the thermocline, superimposed upon the internal, standing wave. These higher frequency waves typically have periods of about a hundred seconds, wavelengths of 10–50m, and amplitudes of 0.05–2m. Most importantly, the waves can roll up and break, just like waves at the surface, opening a temporary window in the thermocline barrier that allows the exchange of heat, nutrients, oxygen, and particles. These billowing effects in a layered fluid are called 'Kelvin–Helmholtz instabilities' (Lord Kelvin yet again, this time in the company of the famous German physicist, Hermann von Helmholtz), and they can sometimes be observed in the sky as swirls of cloud that form as warm air is mixed into cold. The onset of these billows depends on the velocity gradient across thermocline. They can occur near the lake margin, giving rise to increased nutrient supply and primary productivity there, or this mixing can occur mid-lake, causing lake-wide algal growth.

Currents in the lake

The oscillating flows associated with surface and internal seiches are only a subset of the dazzling variety of water mass movements that occur within lakes. Forel pointed out that at the largest dimension, at the scale of the entire lake, there has to be a net flow from the inflowing rivers to the outflow, and he suggested that from this landscape perspective, lakes might be thought of as enlarged rivers. Of course, this riverine flow is constantly disrupted by wind-induced movements of the water. When the wind blows across the surface, it drags the surface water with it to generate a downwind flow, and this has to be balanced by a return movement of water at depth. This explains the observations by the Lake Geneva fisherman that their deep nets could be displaced in a direction that was opposite to the prevailing wind.

In large lakes, the rotation of the Earth has plenty of time to exert its weak effect as the water moves from one side of the lake to the other. As a result, the surface water no longer flows in a straight line, but rather is directed into two or more circular patterns or gyres that can move nearshore water masses rapidly into the centre of the lake and vice-versa. Gyres can therefore be of great consequence, short-circuiting water and its associated content, including pollutants and even toxic algae, from one location to another. They give rise to a striking pattern of water flow; for example, at Lake Biwa, Japan, which is said to have the most beautiful gyres in the world (Figure 15). Unrelated to the Coriolis Effect, the interaction between wind-induced currents and the shoreline can also cause water to flow in circular, individual gyres, even in smaller lakes.

At a much smaller scale, the blowing of wind across a lake can give rise to downwind spiral motions in the water, called 'Langmuir cells'. These were first observed by the distinguished American

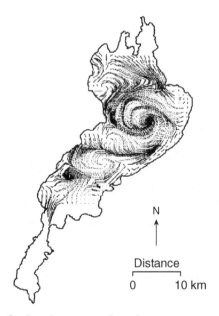

15. The gyres of Lake Biwa, Japan, based on measurements with an Acoustic Doppler Current Profiler (ADCP) on board the research vessel Hakkengo.

scientist Irving Langmuir in the Sargasso Sea, where adjacent, counter-rotating spirals of water movement concentrate floating material at the surface. In that region of the ocean, the material is seaweed of the genus *Sargassum*, which forms long parallel lines of accumulation. These circulation features are commonly observed in lakes, where the spirals progressing in the general direction of the wind concentrate foam (on days of white-cap waves) or glossy, oily materials (on less windy days) into regularly spaced lines that are parallel to the direction of the wind.

Density currents must also be included in this brief discussion of water movement, for they play a major role in many if not most of the world's lakes, including Lake Geneva. Just like stratification, this feature is a result of the density–temperature relationship of water. Cold river water entering a warm lake will be denser than its surroundings and therefore sinks to the bottom, where it may continue to flow for considerable distances. In Lake Geneva, the cold, sediment-laden water of the upper Rhône River enters the lake, immediately plunges and moves many kilometres along the lake floor where it has cut a deep ravine called the Rhône Canyon. Measurements in that canyon using multi-beam echo-sounders indicate that in some places the sediments can be eroded by several metres each year by this process, while in other locations the transported sediment builds up as bottom deposits. In this way, the submarine canyon is an ever-changing, sinuous valley at the bottom of the lake, and a conduit and etching track for its glacial inflow. Density currents contribute greatly to inshore–offshore exchanges of water, with potential effects on primary productivity, deep-water oxygenation, and the dispersion of pollutants.

Chapter 4
Life support systems

> No serious analysis, to my knowledge, has indicated any lake water completely free of microbes...
>
> F. A. Forel

François Forel was on the right track when he acknowledged the widespread presence of microbes, but little could he suspect the astonishing variety and abundance of microscopic life that underpins the ecology of natural waters. A cup of lake water scooped from even the clearest of lakes will contain a populous yet invisible living world: perhaps 100,000 photosynthetic cells, ten million bacterial cells, and a hundred million 'wild viruses' in suspension, all unapparent to the naked eye. Freshwater ecologists have long puzzled over the co-existence of so many different types of algal cells in the homogenous surface waters of lakes. G. Evelyn Hutchinson, one of the most renowned of lake scientists, called this the 'paradox of the plankton': why is it that one species does not simply drive all others to extinction by outcompeting them for the limited resources? With the advent of new molecular and biochemical techniques, this spectacular diversity of microbes has become even more apparent. Just as we now realize that the ensemble of microbes that live on and within our bodies, the 'human microbiome', can greatly affect our state of health, the 'aquatic microbiome' is central to the healthy functioning of lake ecosystems and their responses to environmental change.

Solar-based economies

Almost all ecosystems on Earth depend on the input of energy from the sun, either directly for photosynthesis in the present, or indirectly through the past accumulation of photosynthetic biomass, and its subsequent use by microbes and animals. The recycling of old plant material is especially important in lakes, and one way to appreciate its significance is to measure the concentration of CO_2, an end product of decomposition, in the surface waters. This value is often above, sometimes well above, the value to be expected from equilibration of this gas with the overlying air, meaning that many lakes are net producers of CO_2 and that they emit this greenhouse gas to the atmosphere. How can that be?

To find an answer to this question we need to move outside the boundary of the water-filled basin. Lakes are not sealed

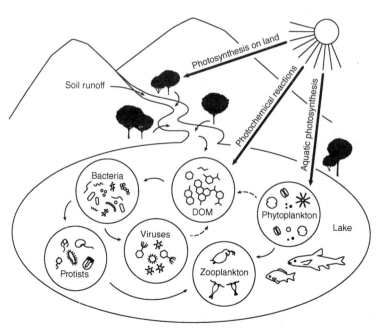

16. From sunlight to diverse microbes and the aquatic food web.

microcosms that function as stand-alone entities; on the contrary, they are embedded in a landscape and are intimately coupled to their terrestrial surroundings. Organic materials are produced within the lake by the phytoplankton, photosynthetic cells that are suspended in the water and that fix CO_2, release oxygen (O_2), and produce biomass at the base of the aquatic food web. Photosynthesis also takes place by attached algae (the periphyton) and submerged water plants (aquatic macrophytes) that occur at the edge of the lake where enough sunlight reaches the bottom to allow their growth. But additionally, lakes are the downstream recipients of terrestrial runoff from their catchments (Figure 16). These continuous inputs include not only water, but also subsidies of plant and soil organic carbon that are washed into the lake via streams, rivers, groundwater, and overland flows.

The organic carbon entering lakes from the catchment is referred to as 'allochthonous', meaning coming from the outside, and it tends to be relatively old because it was produced by plants on land in the past. In contrast, much younger organic carbon is available for microbes and the food web as a result of recent photosynthesis by the phytoplankton and littoral communities; this carbon is called 'autochthonous', meaning that it is produced within the lake. This young, dissolved, organic matter is mostly composed of small molecules and it is rapidly consumed by lake bacteria as a preferred source of carbon and energy. On the other hand, much of the allochthonous organic matter is composed of humic and fulvic acids that are derived from terrestrial plant material. These tea-coloured acids are large polymers composed of organic carbon rings (Figure 16). They are highly resistant to decomposition by most bacteria, although they are subject to decay by certain groups of fungi.

It used to be thought that most of the dissolved organic matter (DOM) entering lakes, especially the coloured fraction, was unreactive and that it would transit the lake to ultimately leave unchanged at the outflow. However, many experiments and field observations have shown that this coloured material can be partially broken down by sunlight. These photochemical reactions result in the production of CO_2, and also the degradation of some of the organic polymers into smaller organic molecules; these in turn are used by bacteria and decomposed to CO_2. This sunlight-driven chemistry begins in the rivers, and continues in the surface waters of the lake. Additional chemical and microbial reactions in the soil also break down organic materials and release CO_2 into the runoff and ground waters, further contributing to the high concentrations in lake water and its emission to the atmosphere. In algal-rich 'eutrophic' lakes there may be sufficient photosynthesis to cause the drawdown of CO_2 to concentrations below equilibrium with the air, resulting in the reverse flux of this gas, from the atmosphere into the surface waters.

Precarious oxygen

There is a precarious balance in lakes between oxygen gains and losses, despite the seemingly limitless quantities in the overlying atmosphere. This balance can sometimes tip to deficits that send a lake into oxygen bankruptcy, with the O_2 mostly or even completely consumed. Waters that have O_2 concentrations below 2mg/L are referred to as 'hypoxic', and will be avoided by most fish species, while waters in which there is a complete absence of oxygen are called 'anoxic' and are mostly the domain for specialized, hardy microbes. In the warm waters of the Amazon, for example, the oxygen balance can shift to anoxia during the night, and the fish must therefore migrate into shallower oxygenated waters or out to the river—where predatory species such as large catfish are lying in wait. In many temperate lakes, mixing in spring and again in autumn are the critical periods of re-oxygenation from the overlying atmosphere. In summer, however, the thermocline greatly slows down that oxygen transfer from air to deep water, and in cooler climates, winter ice-cover acts as another barrier to oxygenation. In both of these seasons, the oxygen absorbed into the water during earlier periods of mixing may be rapidly consumed, leading to anoxic conditions.

Part of the reason that lakes are continuously on the brink of anoxia is that only limited quantities of oxygen can be stored in water because of its low solubility. The concentration of oxygen in the air is 209 millilitres per litre (ml/L; 20.9 per cent by volume), but cold water in equilibrium with the atmosphere contains only 9ml/L (0.9 per cent). This scarcity of oxygen worsens with increasing temperature (from 4°C to 30°C the solubility of oxygen falls by 43 per cent), and it is compounded by faster rates of bacterial decomposition in warmer waters and thus a higher respiratory demand for oxygen. Additionally, the microbial demand for this sparse resource can be excessive because lakes are intense sites of decomposition for the entire landscape, with large populations of oxygen-consuming bacteria that are fuelled by organic carbon from the watershed as well as within-lake sources (Figure 16).

Identifying the invisible

In his description of the microbiology of Lake Geneva, Forel in 1904 pointed out that although bacteria were abundant everywhere, they were not to be feared: 'the immense majority of these tiny life forms are entirely innocent'. Innocent perhaps, but certainly not inconsequential given their massive population size as well as diversity of species and job descriptions. Up until recently, identifying the bacteria present and their specific roles in lake ecosystems was simply impossible because most species could not be brought into culture, and most cannot be identified or differentiated by just looking at them under a microscope. These days, however, the application of methods based on nucleic acid sequencing, specifically DNA to examine the genes present in the community and RNA to determine what genes are being read or 'expressed' to produce proteins, is allowing lake samples to be analysed without going through the problematical culture step.

These analyses are revealing an extraordinarily diverse community of active microbiota and microbial functions that collectively make up the lake microbiome. Many groups and species are still poorly understood, and new microbes and processes continue to be discovered every year.

Microbiomes are composed of four microbial components, and there is immense species and functional diversity within each of them. The smallest, most abundant, and probably most diverse are viruses. These miniaturized parasites attack and reprogramme cells to produce more viral particles, and they typically range in size from 20 to 200 nanometres (nm—one millionth of a millimetre). Each cellular microbe has its own set of pathogenic viruses, and because bacteria are the most abundant cells in the microbiome, the majority of naturally occurring 'wild viruses' in the water are bacterial parasites, called 'bacteriophages' or simply 'phages'. Others, however, attack various other components of the microbial food web, which may affect seasonal variations in this food web to an extent that is presently not well understood.
A group called 'giant viruses' (Mimivirus and relatives) has attracted great interest because of their size (greater than 250nm) and their role as parasites of amoebae and algae. There are also viruses that attack aquatic animals, such as the fish parasite 'infectious haematopoietic necrosis virus' (IHNV). This infects trout and salmon species and can cause high rates of mortality in fish farms.

Once the progeny of viruses are fully assembled during the reproduction phase, the host cell is then induced to 'lyse' (i.e. burst open) and release them, in the process disgorging other cellular materials into the water. These materials are choice substrates for uptake and growth by bacteria that have so far eluded attack, but this success may be short-lived before these other bacteria are themselves infected by viruses, or are eaten by protists, who in turn are eaten by zooplankton (Figure 16). This shunting of organic carbon from bacteria via viral lysis to other bacteria then to protists is referred to as the 'viral shunt', and in some lake and ocean environments it may account for more than 10 per cent of the total carbon flow. Viruses play an additional role in transferring chunks of DNA among hosts, which may confer new or modified gene functions that can then be passed onto future generations, although only if the host is lucky enough to avoid viral lysis in this continuous microbial warfare.

The second component of the microbiome is the bacteria themselves, and many branches (phyla) of the bacterial 'tree of life' are well represented in lakes. One way to appreciate their abundance and diversity is to dye the cells in a sample of lake water with a fluorescent stain, filter the cells onto a membrane, and look at the membrane under a fluorescence microscope. The sample will light up like the Milky Way, with thousands of star-bright, fluorescing cells that vary in size and form: mostly spheres (i.e. cocci) but also oblong (i.e. rods), spiral, kidney-shaped, and filamentous forms. It will be seen that most of the cells in this microbial constellation are extremely small, in the range 200–400nm in diameter, too small for Forel to have detected with his standard microscope. These so-called 'ultramicrobacteria' have the advantage of exposing a large cell surface, relative to their volume, to the lake environment. This maximizes their chances of encountering and absorbing (through specialized 'transport' proteins on their outer membranes) the organic molecules and nutrients that occur at dilute concentrations in the lake water.

The most common phylum of pelagic bacteria is the Proteobacteria, with three notable subphyla in lakes; alpha-, beta-, and gammaproteobacteria. Betaproteobacteria are the most abundant, constituting up to 70 per cent of the total number of cells in the plankton. They include the aptly named lake inhabitant, *Limnohabitans*, which appears to be able to grow rapidly on organic materials released by phytoplankton and thereby outpace the dual pressure of grazing and viral attack. Another common betaproteobacterium found throughout the world's lakes is

Polynucleobacter, which has the capacity to utilize complex organic materials, including breakdown products of humic acids flowing in from the catchment. A betaproteobacterium that plays an especially important role in the nitrogen cycle is the nitrifier *Nitrosomonas*, which oxidizes ammonium (NH_4^+) to nitrite and in the process consumes large quantities of oxygen.

Gammaproteobacteria is a bacterial group that mostly occurs in marine waters, but there are two subgroups that deserve special attention in lakes. The family Methylococcaceae includes several genera that use methane as their carbon and energy source; these include *Methanococcus* and *Methylobacter*, often found near anoxic environments where methane is being generated, such as on the surface of lake sediments. Another family is the Enterobacteriaceae, and its most infamous member is *Escherichia coli*. This species, usually abbreviated as *E. coli*, is named after the Austrian pediatrician Theodor Escherich who first isolated it from the faeces of an infant. Most *E. coli* are not pathogenic (although there are dangerous exceptions), but they are used in monitoring of drinking water and bathing areas on lakes as an indicator of human faecal contamination. Apart from infectious strains of *E. coli*, this contamination may include other pathogenic microbes causing water-borne diseases such as cholera, hepatitis, typhoid, and gastrointestinal illnesses.

Most of the bacterial species in lakes are decomposers that convert organic matter into mineral end products, in particular carbon dioxide, ammonium, phosphate, and hydrogen sulfide (H_2S). In addition to this essential composting and recycling role, some bacteria specialize in transforming inorganic molecules as a source of energy (such as the nitrifiers), and there are others that depend upon sunlight. The most notable of the latter are apparent under the fluorescent microscope as brightly fluorescing red-orange balls amongst the constellation of stained cells. These are 'picocyanobacteria' and although they are larger than the decomposers, they are still pretty small, around 2 micrometres (μm) or less. They would have escaped detection by Forel with a standard microscope, yet they are probably the most abundant photosynthetic cells in Lake Geneva, as in most lakes and oceans. Their red-orange glow in a fluorescence microscope is due to their blue and red protein pigments that, along with chlorophyll, absorb light and fluoresce.

Archaea (or archaeons), the third constituent of microbiomes, share certain features with bacteria in that their cells are small, non-descript, and 'prokaryotic'; that is, they lack a nucleus and other cellular structures that are typical of more advanced 'eukaryotic' cells, including our own. This simplicity, however, belies a set of unusual features that makes them genetically and biochemically distinct, and microbiologists classify archaeons apart from bacteria and eukaryotes as the 'third domain of life'. Some of these microbes play important biogeochemical roles in natural waters, such as the production of methane and the oxidation of ammonium.

The last but certainly not least component of the lake microbiome is the assemblage of 'microbial eukaryotes' with their more complex nucleated cells. Also called 'protists', these include two major groups that were historically separated by function: photosynthetic protists or algae, which capture sunlight for photosynthesis and use CO_2 as their carbon source; and colourless protists that are fuelled by organic molecules that they absorb from the lake water or extract from bacteria. Even the healthiest of algae leak some of their products of photosynthesis into the water during their growth and reproduction, and much more of this organic matter is released when the algal cells are broken up by zooplankton or burst by viruses. This organic carbon would be lost from the food web were it not for its uptake by bacteria and then capture of those bacteria by protists. This carbon recovery process is referred to as the 'microbial loop', and some of this carbon and energy can then move up the food web via zooplankton, ultimately to fish (Figure 16).

Not so very long ago, biologists had a clearly defined view of the living world that was based on carbon and energy sources: inorganic versus organic, photosynthesis versus feeding, plants versus animals. However, protists have little respect for this scientific clarity because many are able to alternate between plant and animal modes of life. Such species that depend upon this mixed combination of energy sources are referred to as 'mixotrophs', and they occur commonly in most lake waters. This strategy allows them to exploit the best of both worlds by harnessing sunlight with their photosynthetic pigments and by exploiting the local resources of pre-made organic compounds. Their capture of bacteria is especially effective, for these minute cells have already done the hard work of scavenging organic molecules and nutrients from the surrounding lake water. The bacterial cells thereby provide concentrated, high-energy food packages for mixotrophic protists, as well as for non-photosynthetic protists such as colourless flagellates and ciliates.

Cycles that matter

Lake microbiomes play multiple roles in food webs as producers, parasites, and consumers, and as steps into the animal food chain (Figure 16). These diverse communities of microbes additionally hold centre stage in the vital recycling of elements within the lake ecosystem, in particular by their activities of oxidation (loss of electrons) and reduction (gain of electrons). These biogeochemical processes are not simply of academic interest; they totally alter the nutritional value, mobility, and even toxicity of elements. For example, sulfate is the most oxidized and also most abundant form of sulfur in natural waters, and it is the ion taken up by phytoplankton and aquatic plants to meet their biochemical needs for this element. These photosynthetic organisms reduce the sulfate to organic sulfur compounds, and once they die and decompose, bacteria convert these compounds to the rotten-egg smelling gas, H_2S, which is toxic to most aquatic life. In anoxic waters and sediments, this effect is amplified by bacterial sulfate reducers that directly convert sulfate to H_2S. Fortunately another group of bacteria, sulfur oxidizers, can use H_2S as a chemical energy source, and in oxygenated waters they convert this reduced sulfur back to its benign, oxidized, sulfate form.

The carbon cycle is at the heart of ecosystem functioning (Figure 17). Microbes are responsible for many of the key transformations, but mineral chemistry also plays a major role. Inorganic carbon enters the lake in three different forms: gaseous CO_2, bicarbonate ions (HCO_3^-), and carbonate ions (CO_3^{2-}). Carbonate is mainly derived from the weathering of limestone (calcium carbonate) and dolomite (calcium and magnesium carbonate) in the watershed, while the CO_2 enters from the atmosphere, inflows, and from respiration, mostly associated with the bacterial decomposition of organic matter. These three forms are in chemical equilibrium with each other via the equation:

$$2H^+ + CO_3^{2-} \rightleftharpoons H^+ + HCO_3^- \rightleftharpoons H_2CO_3 \rightleftharpoons CO_2 + H_2O$$

This equation is central to the pH balance of the lake, because it means that acids entering the water will be rapidly neutralized: the carbonate and bicarbonate ions will take up acid protons (H^+). However, this acid neutralizing capacity (or 'alkalinity') varies greatly among lakes. Many lakes in Europe, North America, and Asia have been dangerously shifted towards a low pH because they lacked sufficient carbonate to buffer the continuous inputs of acid rain that resulted from industrial pollution of the atmosphere. The acid conditions have negative effects on aquatic animals, including by causing a shift in aluminium to its more soluble and toxic form Al^{3+}. Fortunately, these industrial emissions have been regulated and reduced in most of the developed world, although there are still legacy effects of acid rain that have resulted in a long-term depletion of carbonates and associated calcium in certain watersheds.

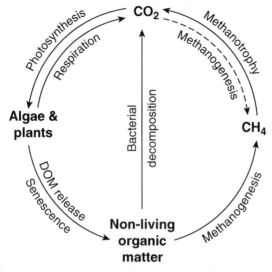

17. The aquatic carbon cycle.

CO_2 is continuously removed by photosynthetic microbes, the phytoplankton, as well as by water plants. To make up for that shortfall and maintain equilibrium, the inorganic carbon reactions are driven to the right to replace the consumed CO_2, and in the process they consume protons and thereby cause an increase in pH. At higher pH values and especially in warm water, the carbonate is no longer soluble and it precipitates out as a chalky coloured suspension. Astronauts on board the International Space Station have taken some striking pictures of such events, called 'whitings', in the North American Great Lakes.

Methane (CH_4) is the second gas of great interest in the aquatic carbon cycle (Figure 17), especially because it is a greenhouse gas like CO_2, but with more than twenty times the greenhouse warming potential per molecule. The generation of methane (i.e. methanogenesis) primarily takes place in anoxic environments by a variety of specialized archaea, the third domain of cellular life in the lake microbiome. Some of these microbes use CO_2 (the dashed line in Figure 17), while others use small organic molecules to produce the highly reduced CH_4. Mostly this takes place in the black-ooze sediments at the bottom of organic rich lakes, but it may also occur in bottom waters that have lost their oxygen, or in both the water and sediments within ice-covered lakes that become completely anoxic in winter.

Some impressive examples of winter gas production can be found in the thermokarst lakes that occur in great abundance across the Arctic tundra. These ecosystems are enriched by soil organic carbon that enters the lakes from thawing and eroding permafrost. The bacterial decomposition of that carbon causes rapid oxygen loss soon after freeze-up, and these ice-capped anoxic waters are ideal bioreactors for methane production by archaea. Making a hole through the ice will release methane bubbles that have accumulated over winter, and this vented gas can be set alight to produce a spectacular burst of flame above the snow and lake ice.

Organic molecules contain carbon in its reduced form, and the most reduced of all is methane. Oxidation back to CO_2 completes the cycle (Figure 17). In the case of most of the particulate and dissolved organic matter, derived mostly from dead algae and plants, this composting role is undertaken by many species of bacteria that can break down and derive energy from a great variety of organic substrates. For methane oxidation, this is the preserve of a more limited number of bacterial specialists called 'methanotrophs'. These bacteria are normally confined to a narrow interface zone where there is both methane and oxygen. A notable exception is in thermokarst lakes, where re-oxygenation from the air after the ice melts in summer offers a paradise of methane plus oxygen for methanotrophs. As might be expected, these microbes make up an unusually large fraction, sometimes more than 10 per cent, of all the bacteria in these permafrost thaw waters during summer.

Nitrogen cycling has a much greater complexity relative to the carbon cycle in terms of the different types of molecules and ions, their states of oxidation, and the array of microbial specialists that keep the cycle spinning (Figure 18). Nitrogen is the most abundant gas in the atmosphere, and also in lake water, but the triple bond of the N_2 molecule is extremely stable and difficult to break. Some cyanobacteria are capable of this enzymatic feat, notably certain planktonic bloom-formers such as *Dolichospermum* (previously called *Anabaena*) and bottom-dwelling forms such as *Nostoc*, which produces jelly-like sheets and balls. In general, however, N_2-fixation is not a major source of nitrogen for lake ecosystems. For example, in Lake Mendota, Wisconsin, one of the most studied lakes in the world, beginning with the classic work of pioneer limnologists Edward A. Birge and Chancey Juday, blooms of cyanobacteria occur each year, including nitrogen-fixing species. However, this biological fixation of atmospheric N_2 appears to account for less than 10 per cent of all nitrogen entering the lake.

Most of the nitrogen entering Lake Mendota, and lakes in general, is derived from their watersheds via inflows, and from their airsheds (the local or regional atmosphere) in rain, snow, and

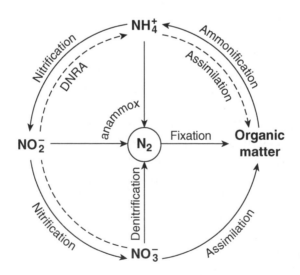

18. The aquatic nitrogen cycle.

wind-blown dust. This nitrogen arrives as nitrate, ammonium, and a variety dissolved and particulate organic forms. Some of this nitrogenous material is taken up by the phytoplankton during their cell growth (the assimilation processes in Figure 18), and when the cells die, some of this organic matter is eventually decomposed to organic nitrogen and ammonium ('ammonification').

Several processes lead the cycle back to nitrogen gas. Apart from being the preferred nitrogen for uptake by phytoplankton, ammonium (NH_4^+) can also be oxidized by a group of specialized bacteria (and some archaea) to nitrite (NO_2^-), and by other bacteria then to nitrate (NO_3^-). Some nitrifiers called 'complete ammonium oxidizers' (abbreviated comammox) can convert NH_4^+ all the way through to NO_3^-. As indicated by these chemical formulae, nitrifying bacteria have a gourmand appetite for oxygen: three atoms of oxygen are needed for each atom of nitrogen to be fully oxidized to nitrate. In the bottom waters of eutrophic lakes, the precarious oxygen balance can be tipped to complete anoxia in part because of this extremely large demand.

In anoxic conditions, other bacteria take the lead in the ongoing transformation of nitrogen. Certain species convert nitrate to ammonium, a process called nitrate ammonification or, in even more longwinded terms, dissimilatory nitrate reduction to ammonium (DNRA in Figure 18). Another bacterial group of great importance to lakes converts the nitrate to nitrogen gas, which is then lost to the atmosphere. This process of denitrification thereby reduces the total amount of nitrogen remaining in the lake ecosystem. Yet one more group of specialists on the N-cycle list is that of anammox (anaerobic ammonium oxidizing) bacteria. These combine nitrification with denitrification, and in the process they produce N_2 gas. There is much interest in using these bacteria (members of the phylum

Planctomycetes) for wastewater treatment because they can convert nitrogen-containing waste to N_2 that is vented to the air, and they have little requirement for organic carbon supplements, unlike denitrifying bacteria.

The biogeochemical phosphorus cycle comprises another set of oxidation and reduction processes that are immensely important for lake ecosystems. This element is often a limiting factor for algal growth, and is a major cause of lake enrichment by domestic wastes and fertilizers from agriculture. Unlike carbon and nitrogen, phosphorus has no gaseous form, and it arrives from catchment rocks and soils as suspended particulate matter and in dissolved forms such as orthophosphate and dissolved organic phosphorus. Phosphorus is best measured as 'total phosphorus' (TP), the sum of all dissolved and particulate forms, and TP concentrations range from less than 10 parts per billion (ppb) in clear oligotrophic lakes to 100ppb in algal-rich, eutrophic lakes.

Huge quantities of phosphorus are stored in lake sediments, but while most of this is locked away from the overlying water, some of it can be mobilized and released under the right conditions. This was first demonstrated by the eminent limnologist Clifford H. Mortimer in a classic experiment using sediments from the bottom of Lake Windermere, in the English Lake District. He placed this mud on the bottom of an aquarium, overlaid the mud with lake water, and then measured the chemical changes associated with oxygen loss. Once the oxygen was depleted, there was an outpouring of dissolved iron and phosphate from the mud into the overlying water. This same effect can now be demonstrated with much greater resolution using micro-electrodes. In an experiment with sediments from Lake Erie, the shallowest waterbody of the North American Great Lakes, profiles with a novel phosphate electrode showed a rise in concentration in the surface sediments and in the overlying water by a factor of 10,000 after the shift from oxygenated to anoxic conditions (Figure 19).

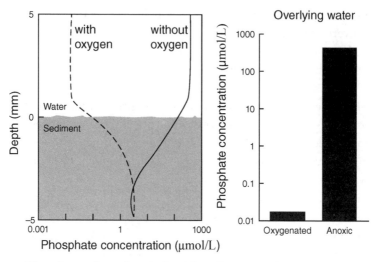

19. **Phosphate release from Lake Erie sediments under oxygenated and anoxic (without oxygen) conditions.**

Several mechanisms are responsible for these chemical changes at the sediment surface. Mortimer correctly surmised that much of the phosphorus in oxygenated sediments is attached (adsorbed) to insoluble iron, or ferric (Fe III) hydroxy-oxides. Under anoxic conditions these oxides are reduced to the ferrous (Fe II) form, which is soluble, and this allows the phosphate to be released into the overlying water. Furthermore, bacterial reduction of sulfate during anoxia produces sulfide that reacts and binds with the iron, resulting in less ferric hydroxy-oxide for phosphate adsorption. Additionally, oxygenated sediments may have a surface cap of bacteria that sequester phosphorus internally as polyphosphate granules, which may be released under anoxic conditions. There are other variables such as aluminium, calcium, organic carbon, and pH that strongly influence this complex biogeochemistry, and not surprisingly, there are large differences in the magnitude of this effect among lakes, including no response to anoxia in some. In many eutrophic lakes, however, once this oxic–anoxic threshold is crossed (Figure 19), phosphorus that was derived from the catchment over years to decades earlier is no longer locked in the sediments, and its release can accelerate enrichment and slow down attempts to improve lake water quality.

Zones of production

With the methods available to him at the time, it was impossible for Forel to be aware of the diversity of teeming microbes that support the natural cycles of lakes, and only now are we beginning to realize the genetic richness of these species and the complex network relationships among them. Forel was aware, however, that lakes can be divided into two zones that sharply differ in terms of their photosynthetic communities at the base of the aquatic food web. In the inshore (or 'littoral') zone, the primary producers are mostly aquatic plants, some rooted in the sediments and fully submerged, and others with leaves that float or extend out of the water. In shallow lakes and ponds, these water plants or 'aquatic macrophytes' and their associated microbes may dominate the overall biological production of the ecosystem. Forel was impressed by the luxuriant growth of these plant communities in Lake Geneva, noting in 1904 how 'they form true underwater forests, as picturesque, mysterious and attractive as the most beautiful forests of our mountains'. These provide important habitats for aquatic animals, as well as trapping nutrients and affecting water currents.

Offshore, in the limnetic or pelagic zone, beauty is more at the microscopic level, and the sweep of a plankton net through the water will yield a wonderfully diverse collection of pigmented cells that vary in size, shape, and colour. This algal plankton or phytoplankton captures light for photosynthesis, and their growth extends down to the bottom of the photic zone. This also demarcates the bottom and therefore offshore extent of the littoral zone, although individual plant species may be limited in their distribution by other factors such as water pressure, grazing animals such as crayfish, and the type of substrate. At greater depths in the lake lies the profundal zone, where the organisms live in perpetual darkness and depend on organic materials arriving from above, especially the continuous rain of phytoplankton cells sinking out of the photic zone (see Chapter 5).

The phytoplankton in most lakes contains dozens if not hundreds of species, and these fall into four major groups. First and foremost are the diatoms, with their highly ornamented walls of silica glass. These algal cells can be observed and identified with a standard microscope, and Forel appreciated their importance in the Lake Geneva food web, writing in 1904:

> a diatom, is eaten by a rotifer, which is eaten by a copepod, which is eaten by a cladoceran, which is eaten by a whitefish, which is eaten by a pike, which is eaten by an otter or by a human.

Glass is a heavy substance, and diatoms are therefore at the mercy of gravity, many of them sinking out of the photic zone and accumulating in the sediments. The best time of year for their growth is during spring and autumn, when the full mixing of the lake keeps these heavy cells up in suspension, and when there is adequate sunlight and nutrients for their photosynthetic production. There are long-term diatom records for many lakes of the world, including Lake Windermere, that show the regular rise and fall of diatoms every year. Their seasonal collapse is typically due to the onset of stratification, and the cessation of full mixing. Their rapid demise through sinking may also be hastened by grazing by zooplankton, or by parasitic attack.

Non-swimming green algae are also common in the phytoplankton, and vary enormously in size and shape. The smallest are less than 2 or 3μm in size and are referred to as 'picoeukaryotes'. These can be present in high abundance, but usually require DNA techniques to identify. At the other extreme, some species of green algae form large colonies of cells. For example, a beautiful lace-like species found only at Lake Biwa, Japan, *Pediastrum biwae* variety *triangulatum*, has a colony diameter of almost a tenth of a millimetre. A common species found in many lakes of the world, the gelatinous colonies of *Sphaerocystis schroeteri*, is of a similar or larger size. These dimensions make the particles too big for zooplankton to readily feed on, and large colony size is an effective defence against grazers.

The third group of phytoplankton comprises photosynthetic species that can swim: 'phytoflagellates' that belong to many algal phyla and encompass a great variety of species, both in terms of ecology, and pigmentation. These motile cells propel their way through the viscous liquid environment with flagella, some with one large and one very small flagellum such as the golden-brown species *Dinobryon divergens* that produces tree-like colonies of beating cells (Figure 20), others with two equal flagella such as

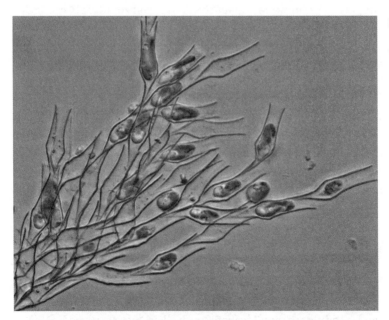

20. The colonial phytoflagellate, *Dinobryon divergens*. The individual cells are 10–15μm in length and are mixotrophic: they feed on bacteria as well as capture sunlight for photosynthesis.

the green alga *Chlamydomonas*. A mammoth among these motile phytoplankton is the brown-coloured dinoflagellate, *Ceratium hirundinella*, found in many lakes throughout the world; the cells are up to a quarter of a millimetre in length (250μm) and are able to swim up and down in the epilimnion each day.

The fourth group is the cyanobacteria, also known by their earlier name 'blue-green algae', with their distinctive combination of green chlorophyll and blue protein pigments. These include picoplankton species (picocyanobacteria) but also large colonial forms such as *Microcystis aeruginosa*. Cyanobacteria in general prefer warm temperatures and their large colonies are especially abundant in late summer and autumn, when they may form dense blooms and create major water quality problems.

The total amount and composition of the phytoplankton provide important information about the biological productivity of a lake as well as its water quality. The most rigorous approach is to analyse a sample using an inverted microscope; a glass-bottomed cylinder of lake water is allowed to sediment and is then examined though inverted lenses in the microscope that look up through the bottom of the cylinder to view the phytoplankton that have settled on the glass. This is a time consuming analysis, and it also requires a high level of skill by the microscopist to differentiate algal cells from detritus, and to identify the species.

Another, complementary approach is to measure the amount of chlorophyll *a*, a pigment that is found in all phytoplankton, including cyanobacteria. As a further estimate of the abundance of phytoplankton, as well as a guide to what groups are present, algal accessory pigments can be measured by high pressure liquid chromatography. In most samples, these analyses will show the presence of light-capturing pigments such as fucoxanthin in diatoms and peridinin in dinoflagellates, as well as diverse pigments that protect the cells against photodamage by bright light, such as lutein in green algae and echinenone in cyanobacteria.

This brings us back to Hutchinson's paradox: how can so many different species co-exist in the microscopic world of plankton? He considered several possible explanations, including the idea that the community might be in a state of disequilibrium; the lake environment is continuously changing, and so one species that is the winner today will be less favoured tomorrow, leading to a mixture of species and insufficient time to completely exclude any losers before their optimal conditions return. This idea has been further developed in the era of genomic analyses, which has revealed how the planktonic diversity of the lake microbiome is so much greater than even Hutchinson could have envisaged. Microbiologists talk about 'the rare biosphere' where most microbial species, including phytoplankton, are in low abundance and growing slowly or not at all for much of the time, while the most abundant species are subject to the heaviest losses by viruses and grazers. At the scale of a lake, even a residual population of a few cells per millilitre translates into a huge lake-wide population, which provides a hedge against extinction, and an inoculum to seize the day during the next round of favourable growth conditions.

Chapter 5
Food chains to fish

> The small and the weak are prey for the large and the strong; they in turn are devoured by the larger and stronger, or, if they escape, they will not avoid the microbes of decomposition that all organisms, directly or indirectly, are subject to.
>
> F. A. Forel

In his autobiography, François Forel recounts how one of his most exciting moments in lake science was discovering animal life at the bottom of Lake Geneva. The first sign came from his analysis of ripples in the sediments, just offshore from the family home at Morges. He had placed a sample of this bottom material under the microscope to determine its composition when suddenly a worm-like creature came into view, thrashing among the mineral particles. He was startled by this unexpected appearance of something so obviously alive, and he immediately began to wonder if the sediments of Lake Geneva could be inhabited by such animals to its greatest depths. If this were the case, then 'the profundal zone is not a desert; there is an abyssal society'.

That night, Forel constructed a dredge to sample deeper sediments in Lake Geneva, and in a set of studies that extended from the next day to the rest of his career, he discovered that a great variety of invertebrate animals occurred in the profundal zone (Figure 21), to the very bottom of the lake at around 300m depth. These bottom-dwelling or so-called 'benthic' communities of animals are dependent on the downward rain of organic materials, especially plankton, sinking into the profundal zone from the overlying waters. Forel called this continuous supply from above 'leftovers from the tables of others', and he realized that the benthic communities 'collect everything that falls to the bottom'. In turn, they are a food supply for other animals, including bottom-feeding fish, while bacterial decomposition in the sediments recycles the organic matter from all sources back into dissolved nutrients. Some four decades later, American ecologist Raymond L. Lindeman noted 'the brilliant exposition of Forel' on food webs, bacterial decomposition, and recycling. In his doctoral studies at Cedar Bog Lake, Minnesota, Lindeman built upon and greatly extended these ideas in a quantitative way to produce the 'trophic-dynamic' concept of energy and carbon flow, with bacteria and detritus in the bottom sediments forming the hub that linked all components of the food web.

Higher up in the water column, the life support system (protists and bacteria) described in Chapter 4 provides the carbon and energy for not only the benthic communities, but also for the animal residents of the pelagic or limnetic zone (Figure 21), the

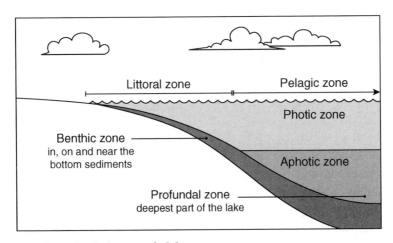

21. **The ecological zones of a lake.**

zooplankton. The largest of these can be readily discerned in a sample of lake water as minute, swimming individuals, 0.2–2mm in size, often moving in discontinuous jumps rather than smoothly gliding in a single direction. These are fed upon by fish that are themselves the prey for others, including larger fish, as well as birds and humans. A great variety of technologies, from high frequency acoustics and satellite telemetry to lipid, isotopic, and genetic analysis, are now combined with the usual methods of observation to better understand the nature of pelagic and benthic food webs and the coupling between them.

Recent studies of aquatic food webs have drawn attention to the importance for many lakes of the role of outside inputs from the surrounding catchment. These are derived from terrestrial plants and soils, and, in this way, they provide subsidies of carbon and energy to meet the needs of the aquatic animals (Figure 16). An additional external influence on lake food webs, but coming from well beyond the limits of the catchment, is the arrival of invasive species. Some of these plants and animals are introduced intentionally to 'improve' the ecosystem. Others accidently make their way into the lake, generally, and with increasing frequency, as a result of human activities like boating and fishing. In many cases, these invaders can severely disrupt the original food web, often to the detriment of the ecosystem services provided by the lake.

Life on the bottom

Forel identified the 'poor worm' moving about in such an agitated fashion on his microscope slide as a species of nematode, or round worm. In fact, lake sediments are the habitat for three important groups of worm-like animals that come from totally different branches (i.e. phyla) of the animal kingdom. Nematodes are the most abundant, and probably the most diverse. These thread-like invertebrates are typically 0.2–2mm in length, and they can occur in concentrations of up to a million per square metre of the lake floor. The smallest individuals are the most abundant, and most live in the upper few millimetres of the sediment. Some 2,000 freshwater species have been described, but the group is not well studied and it is estimated that many thousand more await discovery. They also vary greatly in their feeding strategies; some feed on aquatic plants, protists, or micro-invertebrates, some are larger animal parasites, while others, the dominants in lake sediments, feed on organic particles (detritus), bacteria, or microscopic fungi.

The second group is oligochaetes or segmented worms. Some of these move and burrow through the lake mud and silt, and certain species such as *Peloscolex variegatum* are restricted to well-oxygenated sediments. A major subgroup, the sludge worms, produces a tube of mucilage and particles that is embedded vertically in the sediment; the animals occupy the tube with their heads buried in the sediment for feeding, while their tails protrude through the tube into the overlying water and wiggle about to draw in oxygen. These oligochaetes can be brightly coloured with a red blood pigment that aids their survival in low oxygen environments. Two well-known representatives of this group are *Tubifex tubifex* and *Limnodrilus hoffmeisteri*, which are often found in sediments affected by organic pollution and are indicators of poor water quality.

The third group of worm-like animals in lake sediments is actually the larval form of insects, specifically flies (dipterans). The most common larvae are those of non-biting midges, chironomids (chi- pronounced 'kye' as in kite), and more than 5,000 species are known. These larvae are sought-after food items for benthic fish and other animals, and they can be found in large densities—up to tens of thousands per square metre of lake sediment. Given their much larger body size relative to nematodes, chironomids often dominate the total biomass of benthic communities. Many species have feeding tubes and use their body movement to generate currents that draw in oxygenated water; this burrowing activity as 'ecosystem engineers' can greatly change the oxygen conditions and biogeochemistry of the sediments. Like most other groups of benthic animals, the diversity of species and abundance is often greatest in the littoral zone (Figure 21) because of the diversity of substrates, the presence of plant and algal detritus, and the inputs of organic matter from the catchment. However, in moderate to large size lakes, the area of the littoral zone is small relative to that of the profundal zone, and the total lake biomass of these deep communities may be greater.

Chironomids figure prominently in the history of lake science because they were a favorite research topic for the distinguished zoologist August Thienemann, who was director of the hydrobiological laboratory at Plön, Germany. One of his international colleagues, the aquatic botanist Einar Naumann, had established a field station for the limnological institute at Lund, Sweden, and developed a classification scheme for lakes based on their algal concentrations. He referred to this as the lake's trophic state, from the Greek 'trophikos', meaning nourishment, and he classified waters into the categories of 'oligotrophic' (i.e. with clear water and low phytoplankton abundance) and 'eutrophic' (i.e. rich in phytoplankton). Thienemann adopted Naumann's classification, in common use today, and he showed how totally different assemblages of chironomids occur in these two trophic states; for example, the genus *Tanytarsus* is common in oligotrophic waters while *Chironomus* occurs in eutrophic waters that are low in oxygen. The two lake scientists joined forces in 1922 to found the International Limnological Society (SIL) at an inaugural meeting in Kiel, Germany.

Molluscs are another group of animals that occur in high abundance on the bottom of lakes, with two subgroups: gastropods (i.e. snails) and bivalves (i.e. clams and mussels). Certain fish species prey upon snails, which may find refuge in the littoral zone, on and among the water plants. In this habitat, they graze on detritus and on the biofilms of algae (i.e. periphyton) that coat the plants and the bottom of the littoral zone. Clams are renowned for their long life cycles, with a lifespan of several decades, and for some unionid species, for their impressive dispersal mechanism. These clams have gill membranes that extend outside the shell, in the shape of a small fish (sometimes even with pigmentation suggesting an eye); the membrane undulates with the pumping beat of the clam, and lures in predatory fish. Once the fish bites, the membrane breaks open and releases propagules called 'glochidia'. These latch onto the gills of the unsuspecting fish, and grow as small clams, finally dropping off once they are heavy enough, into the sediments, far away from the site of the parent.

A third group that often constitutes a large fraction of the benthic animal biomass is amphipods, also known as freshwater shrimps or scuds, and they are crustacean scavengers that mostly feed on detritus. Almost 2,000 species live in freshwater, and although a single species may appear to dominate the community, DNA analyses are showing that there are likely to be many hidden or 'cryptic' species that may be morphologically identical but genetically distinct. The greatest adaptive radiation has been observed in Lake Baikal; 260 endemic species have been recognized to date, with an additional eighty subspecies, and hundreds more are likely awaiting detection. Forel observed that an amphipod, 'the blind shrimp', *Niphargus forelii*, was common on sediments throughout the profundal zone of Lake Geneva (Figure 22). The species was subsequently driven to extinction in this lake, but it is still found today in other Swiss, German, and Italian deep alpine waterbodies. In the North American Great Lakes, the amphipod *Diporeia* can account for 50 per cent of the total biomass of benthic invertebrates, and in Lake Biwa, Japan, the endemic species *Jesogammarus annandalei*, can achieve densities up to 63,000/m².

Apart from worms, molluscs, and amphipods, many other animals live in the benthic habitat of lakes, but generally at much lower biomass. These include small species such as rotifers and water mites, and crustaceans in addition to amphipods including

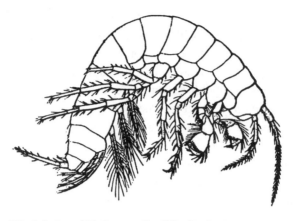

22. The blind shrimp (*Niphargus forelii*) of Lake Geneva.

ostracods (seed shrimps), harpacticoid copepods, and certain cladocerans, especially chydorids. Crayfish occur in a variety of freshwater benthic habitats and often have regional names. For example, they are known by the Maori name koura (*Paranephrops*) in New Zealand lakes, the aboriginal name 'yabby' (*Cherax*) in Australia, and as crawfish (*Procambarus clarkii*) in the southern US, where they are farmed as a sought-after ingredient of Cajun cooking. Crayfish are omnivores, and they mostly keep to the littoral zone where they feed on plant and detrital material, snails, chironomids, and mayflies; in turn they are eaten by fish and birds. Freshwater sponges occur in many lakes, and the polyp stage of freshwater jellyfish may also be found in the benthos, attached to underwater plants and other substrates. As Forel discovered to his wonder, the bottom of lakes, even the deepest of lakes, is most certainly 'not a desert' but rather a zone of biological richness and animal productivity.

Planktonic webs of interaction

In the open waters of the lake, three groups of zooplankton play a leading role in the transfer of carbon and energy from the base of the food web (i.e. phytoplankton and bacteria) to pelagic fish: rotifers, cladocerans, and copepods. The first of these groups is placed in its own branch of animal life, Phylum Rotifera, and is so named because of the rotating, wheel-like appearance of a double crown ('corona') of thread-like cilia that propels it through the water and directs a stream of food particles towards its mouth. These animals were first discovered in a drop of pond water by pioneer microscopist Antonie van Leeuwenhoek, who named them 'wheel animalcules'. They are rare in the sea, but in terms of numbers they are often the most abundant zooplankton in freshwaters. In thermokarst lakes across the northern tundra, for example, they can achieve densities up to 1,500 animals per litre. Rotifers are typically small in size (< 0.2mm), with fast generation times, often just a few days. Most feed on micro-algae, other protists, and bacteria, but there are also carnivorous species, notably in the genus *Asplanchna*. These animals in turn are eaten by copepods and young fish.

The second zooplankton group is cladocerans, which are crustaceans in the size range 0.5–2mm. There are eighty-six genera, of which only four occur in sea. Three of the most common planktonic genera are *Daphnia* (also known as water fleas, although they are not at all parasitic like fleas), *Bosmina*, and *Holopedium*, the latter distinguished by its giant, jelly-containing helmet that covers the head and that may be a defence against predation. Generally, this group is the favourite prey of smaller 'planktivorous' fish. The cladoceran group known as 'chydorids' (family Chydoridae) are found especially in the littoral zone associated with aquatic plants and bottom sediments. The body of cladocerans is covered in an exoskeleton of chitin that is shed as they grow and moult, this process occurring more than twenty times in some species. In some very clear lakes, their carapaces may contain dark melanin pigmentation (as in Figure 23) that acts as a sunscreen in protecting the animal and its eggs against the damaging effects of UV radiation.

Cladocerans have multiple pairs of appendages, with each pair specialized for particular functions. The most prominent

23. Photomicrograph of the zooplankton species *Daphnia umbra* from a lake in Finland. Each animal is around 2mm in length.

appendages are large antennae that are used as paddles for swimming and pulling the animals through the water, which is a viscous medium at the scale of the animal. The legs (four to six pairs) have fine hair ('setae'), with even finer hairs called 'setules' that filter particles from the water, including by electrostatic interactions. This food consists of algae, other protists, bacteria, and detritus, and the collected material is taste-tested by another set of antennae, ground up by other appendages (the mandibles), and worked with mucous into ball of matter ('bolus') that is then moved into the mouth or rejected.

Like rotifers, cladocerans can achieve spectacular population growth over short periods of time as a result of 'parthenogenesis' (Figure 24), and it is common to find temporary ponds and pools where clouds of *Daphnia* seem to appear suddenly from nowhere. Most of the populations are females, and these produce eggs asexually (i.e. without fertilization) that develop in the brood

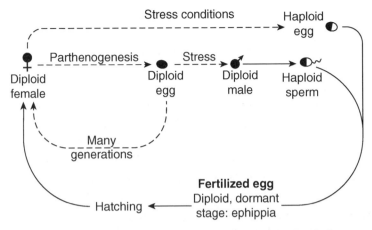

24. Asexual (parthenogenesis) and sexual reproduction by cladoceran zooplankton.

chambers, eventually released as free-swimming 'neonates'. Depending on the species and food conditions, a single mother may carry anything from one egg to more than 200, and in warm water the embryo development time can be as little as two days.

This asexual strategy is extremely effective for rapid growth in stable environments, but cladocerans like some rotifers have the advantage of a sexual mode of reproduction when conditions deteriorate (Figure 24). This could be due to a physical stress such as extreme temperatures, or biological stresses such as crowding and food shortages. At this time the females produce haploid eggs with a single copy of the chromosomes in each egg, and diploid eggs (double copy of each chromosome) that hatch into males. These males then mate with the females and fertilize their haploid eggs to produce a diploid zygote. In many species this is retained with a modified part of the carapace, and is released during moulting as a darkly coloured, encased resting egg called an 'ephippium'. These are highly resistant to extremes such as drying and freezing, and may be important for dispersal between waterbodies in the wind and via bird feathers. They can remain dormant for months, decades, even hundreds of years, for example at the bottom of a desiccated pond, and then hatch into asexual, diploid females once favorable conditions return.

Copepods are among the most abundant zooplankton in the sea, and are also commonly found in almost all lakes. In large deep lakes such as Lake Baikal and the North American Great Lakes, they account for most of the zooplankton biomass. Like cladocerans, copepods are crustaceans with an exoskeleton of chitin and multiple pairs of appendages adapted for swimming, feeding, or sensing (Figure 25). Unlike cladocerans, however, they have no asexual reproductive phase and the populations contain a mixture of males and females. After mating, the female produces eggs that hatch into larvae or 'nauplii', which then moult five or six times

25. Photomicrograph of the copepod *Aglaodiaptomus leptopus* from a lake in southern Quebec, Canada. The animal is 2.3mm long.

before becoming copepodid larvae ('copepodites'). These then undergo five more moults before becoming sexually mature adults. In warm water this entire life cycle may be completed within a week, but in cold lakes of the polar and alpine regions it may take one or more years. Copepods feed on phytoplankton and other protists, and are a lipid-rich food source for planktivorous fish. However, they are harder to catch than are the slower swimming Cladocera, which is why they often dominate in lakes where fish planktivory is strong.

Moving about

Rotifers, cladocerans, and copepods are all planktonic, that is their distribution is strongly affected by currents and mixing processes in the lake. However, they are also swimmers, and can regulate their depth in the water. For the smallest such as rotifers and copepods, this swimming ability is limited, but the larger zooplankton are able to swim over an impressive depth range during the twenty-four-hour 'diel' (i.e. light–dark) cycle. Forel first observed this himself when he rowed out onto Lake Geneva to sample zooplankton at night, and found that his net tow had captured 'myriads of Entomostraca [copepods] that had risen to the surface'. Subsequent work has shown that the cladocerans in Lake Geneva reside in the thermocline region and deep epilimnion during the day, and swim upwards by about 10m during the night, while cyclopoid copepods swim up by 60m, returning to the deep, dark, cold waters of the profundal zone during the day.

Even greater distances up and down the water column are achieved by larger animals. The opossum shrimp, *Mysis* (up to 25mm in length) lives on the bottom of lakes during the day and in Lake Tahoe it swims hundreds of metres up into the surface waters, although not on moon-lit nights. In Lake Baikal, one of the main zooplankton species is the endemic amphipod, *Macrohectopus branickii*, which grows up to 38mm in size. It can form dense swarms at 100–200m depth during the day, but the populations then disperse and rise to the upper waters during the night. These nocturnal migrations connect the pelagic surface waters with the profundal zone in lake ecosystems, and are thought to be an adaptation towards avoiding visual predators, especially pelagic fish, during the day, while accessing food in the surface waters under the cover of nightfall.

One of the most striking effects of fish on migration patterns is for the insect known as the 'phantom midge'. This produces mosquito-like larvae up to 2cm long, also called 'glass worms' because their bodies are transparent apart from a pair of air bags at each end that help them float in the water. In Europe there are two main species that differ in their diel migration behaviour. *Chaoborus flavicans* occurs mainly in fish-containing lakes and ponds, and during the day it remains at the bottom, feeding on animals in the sediments, including in anoxic waters or with its head dug into anoxic sediments where it appears to be capable of anaerobic metabolism based on malate; during the night it migrates to the surface to feed on zooplankton, especially copepods. This migration pattern is most pronounced in the presence of fish, which it appears to be able to detect through chemical signals (fishy-smelling 'kairomones'). The second species, *Chaoborus obscuripes* avoids fish-containing waters, and remains throughout the twenty-four-hour cycle in the near-surface zone, where it is out of the reach of benthic predators such as dragonfly larvae.

Although certain fish species remain within specific zones of the lake, there are others that swim among zones and access multiple habitats. For example, in Lake Superior, the largest freshwater lake by area in the world (82,100km^2; maximum depth 406m), the Cisco (*Coregonus artedi*) is a fish species that is mostly planktivorous, living in the offshore pelagic zone. However, in late autumn it moves inshore to spawn, and its lipid-rich eggs provide an energy subsidy to the food web of the littoral zone, contributing 34 per cent of the energy needs of Lake whitefish (*Coregonus clupeaformis*) that feed mostly on benthic prey such as amphipods in these inshore shallow waters. This type of fish migration means that the different parts of the lake ecosystem are ecologically connected.

For many fish species, moving between habitats extends all the way to the ocean. Anadromous fish migrate out of the lake and swim to the sea each year, and although this movement comes at considerable energetic cost, it has the advantage of access to rich marine food sources, while allowing the young to be raised in the freshwater environment with less exposure to predators. One such example is Arctic char (*Salvelinus alpinus*), a species that is found in deep cold lakes in Great Britain as well as Lake Geneva and elsewhere in Europe. It is the northernmost freshwater fish, and is found all the way to Lake A at latitude 83°N in the Canadian High Arctic. During their freshwater residence, Arctic char feed on benthic invertebrates as well as plankton, small fish, and insects at the water surface; in the sea these animals feed on other fish and amphipods. Genetic markers along with acoustic tags inserted into captured and released fish are now being used to identify distinct populations of this species, and the origin of migrating stocks.

With the converse migration pattern, catadromous fish live in freshwater and spawn in the sea. One example is the European eel, *Anguilla anguilla*, which is most abundant in rivers but also commonly found in natural and artificial lakes. Satellite tagging has shown that adult European eels migrate 5,000km or more to spawn in the Sargasso Sea. At swimming speeds in the range 10–30km per day, this is a long process, taking up to a year and with severe losses by predation. The larvae then return to European waters by the Gulf Stream and North Atlantic Drift.

You are what you eat?

Diet certainly has a major role to play in the nutritional state of animals in the food web of lakes, but changes can also take place via the specific needs and physiology of the species concerned. Take something as simple as the ratio of elements. The famous American oceanographer Alfred C. Redfield established that phytoplankton in the sea typically have a ratio of 106 atoms of carbon to sixteen atoms of nitrogen to one atom of phosphorus (or in terms of weight, 41 grams (g) of carbon to 7g of nitrogen and 1g of phosphorus). He also noted that the same N:P ratio occurred in the regenerated nutrients in the deep ocean. In lakes, the phytoplankton often has a carbon to nitrogen to phosphorus composition that is similar to or slightly above this 'Redfield ratio', but higher up the food web this ratio can differ greatly among animals. The nitrogen to phosphorus ratio for copepods is typically 14g of nitrogen to 1g of phosphorus (i.e. 14:1) or higher, but for cladocerans, even in the same lake, it is often half of that, around 7:1. This strikingly lower ratio in cladocerans has been attributed to their greater quantity of cellular organelles for protein production, specifically ribosomes that contain RNA, a biomolecule that is rich in phosphorus. This allows faster growth rates, but also results in a high biological demand for phosphorus, hence the lower nitrogen to phosphorus ratio. The analysis of nutrient ratios in lakes, 'ecological stoichiometry', has generated important insights and questions in lake science, including about the divergent needs of different animal groups for phosphorus and effects on nutrient recycling.

Food quantity and its rate of supply are important to all animals in the food web, and in general terms, the animal or 'secondary productivity' of lakes increases with increasing phytoplankton and its associated photosynthetic production of biomass or 'primary productivity'. But it is not just about quantity. The nutritional quality of that food can vary greatly, for example in its composition of fatty materials (lipids), and this profoundly affects animal health, reproductive success, and survival. Some of these lipid molecules are brightly coloured and are derived from algal pigments that are transferred to the animals that feed upon them (e.g. the copepod in Figure 25 is bright orange in colour due to the carotenoid astaxanthin). For certain zooplankton, such as in clear alpine lakes, this pigmentation may be primarily a defence against UV radiation, but for other species it appears to be a way to maintain high energy lipid reserves over winter for growth and reproduction the next spring.

Certain lipid molecules called 'polyunsaturated fatty acids' (PUFAs), such as the omega-3 PUFA, eicosapentaenoic acid (EPA), are especially important for the production of hormones that regulate bodily functions such as brain development, vision, and cardiovascular metabolism (including in humans). Most aquatic animals cannot produce PUFAs but must obtain them from their diet, ultimately from the algae that produce them and the consumers of those algae. There is great interest by lake scientists in tracking EPA and related PUFAs through the food web because this provides information on the feeding relationships and health of the aquatic ecosystem. There may even be a transfer of PUFAs to terrestrial species (which tend to be poorer in PUFAs than aquatic organisms) via aquatic insects such as chironomids and dragonflies that are eaten by birds. Lipids are also relevant to understanding the impacts of chemical pollution. Most organic contaminants (e.g. pesticides) are soluble in lipid, and therefore the transfer of lipids through the food web also results in the concentration (i.e. bioamplification) of pollutants in the animals at higher positions (or trophic levels) in the food chain.

One of the most powerful approaches towards analysing food webs is based on naturally occurring isotopes: atoms that have the same number of protons, and are therefore of the same element, but that differ in the number of their neutrons. Nitrogen in the atmosphere, for example, is mostly composed of atoms with seven protons and seven neutrons, denoted 'nitrogen-14' (or, ^{14}N), but a small percentage of it (0.3663 per cent) is composed of nitrogen atoms with an additional neutron, making it nitrogen-15 (^{15}N). When nitrogen is taken up by animals, the ^{15}N is retained to a slightly greater extent than ^{14}N, and this enrichment effect continues step-by-step up the food chain.

These differences based on a single neutron may seem small, but with sensitive mass spectrometers, even minute shifts in ^{15}N enrichment can be accurately detected. In the pelagic zone of Lake Baikal, for example, ^{15}N is enriched by about 3.3ppt, with each step in the food web (Figure 26). Large endemic diatoms in the phytoplankton such as *Aulacoseira baikalensis* take up inorganic

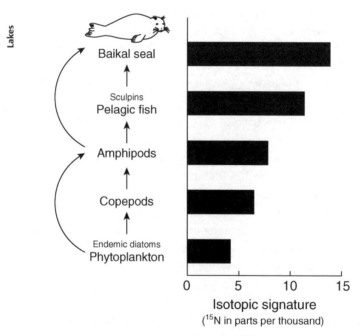

26. **The pelagic food web in Lake Baikal and the corresponding increase in nitrogen-15 ($δ^{15}$N) at each trophic level.**

nitrogen, and their delta ^{15}N (denoted as $\delta^{15}N$ and defined as the difference in the $^{15}N/^{14}N$ ratio relative to the atmosphere) is around 4ppt. The diatoms are eaten by copepods and eventually the nitrogen passes all the way up the food chain, with a final $\delta^{15}N$ value of 14ppt in the seals. There are other sources of variation, but the approach provides a valuable guide to who is eating who in the lake food web, and for omnivorous animals, the relative importance of different food items in their diet. The natural isotopic ratio $^{13}C/^{12}C$ is also used in this way to measure dietary source, and, in marine studies, the sulfur isotopic ratio $^{34}S/^{32}S$ provides a similar tracer. Isotopic fractionation also takes place when water evaporates and passes from liquid to gas, and the isotopic ratios $^{2}H/^{1}H$ and $^{18}O/^{16}O$ in water molecules are used in hydrology to determine the evaporation versus precipitation balance of lakes.

Invaders at the lake

In the late 19th century, Forel observed with alarm the arrival of the Canadian water weed *Elodea canadensis* in Lake Geneva, and its 'exuberant and frightening expansion' throughout the lake. This invasive species had been introduced intentionally in local ponds and streams to improve the fish habitat, and in an unfortunate history repeated in many parts of the world, it soon entered and expanded throughout the littoral environment of the lake. This species and other members of the aquatic plant family Hydrocharitaceae invaded New Zealand lakes in the mid-20th century, forming underwater forests up to 6m tall that greatly altered littoral habitats and interfered with electricity production via their proliferation in hydro-reservoirs. Other species such as the *Myriophyllum spicatum*, the Eurasian water milfoil, is creating problems in drinking water supplies (including at Lake St-Charles, Quebec), and the South American water hyacinth *Eichhornia* is an aggressive floating plant that covers and chokes aquatic habitats in the southern USA, Asia, and Africa, including inshore sites at Lake Victoria.

The arrival of an invasive animal species in a lake can have a massive effect that begins at one level and then propagates throughout the entire food web. A classic example of this 'trophic cascade' was observed in Flathead Lake, a large (500km²), deep (116m) lake in Montana, USA. Over the period 1968 to 1975, the opossum shrimp, *Mysis diluviana* (closely related to the European species, *Mysis relicta)* was introduced into three headwater lakes to improve the salmon fishery. By 1981, the shrimp had drifted downstream to appear in Flathead Lake, where by the late 1980s it had undergone an explosive increase in numbers (Figure 27). Within a few years after the arrival of the mysids at Flathead Lake, there was a collapse of cladocerans and copepods in the zooplankton due to overconsumption by the massive shrimp populations. A 'top-down' food web effect followed, with large increases in phytoplankton biomass and changes in its community composition because of decreased grazing by zooplankton.

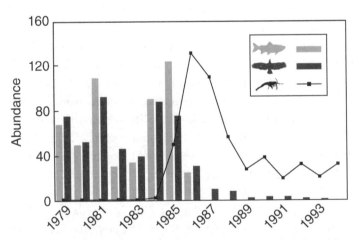

27. Food web changes at Flathead Lake after the invasion of mysid shrimps. The numbers of Kokanee salmon (the plotted values should be multiplied by a hundred) and Bald eagles (multiply by seven) were measured at an upstream salmon spawning site, while the shrimp abundance (multiply by 1,000) was the number of mysids in the water column of the lake per square metre.

The Kokanee salmon, also an introduced species in Flathead Lake, became deprived of their zooplankton food source, and were unable to feed on the mysids, which migrated up into the pelagic zone only at night when the salmon were unable to see them. This unforeseen fish avoidance behaviour has made mysid stocking an especially poor choice for enhancing salmon production. In the Flathead Lake watershed, sport catches of salmon plunged from more than 100,000 through 1985 to zero in 1988 and beyond. Bald eagles, which congregated on the main spawning stream to feed on Kokanee, went from more than 600 in the early 1980s to near-zero a decade later (Figure 27). An additional food web effect was that the mysids became the major food source for another introduced species, bottom-feeding Lake trout (*Salvelinus namaycush*), and the current rise of this species is driving the native Bull trout (*Salvelinus confluentus*) towards extinction.

Invasive species that are the most successful and do the most damage once they enter a lake have a number of features in common: fast growth rates, broad tolerances, the capacity to thrive under high population densities, and an ability to disperse and colonize that is enhanced by human activities. Zebra mussels (*Dreissena polymorpha*) get top marks in each of these categories, and they have proven to be a troublesome invader in many parts of the world. Their native habitat is in the Caspian Sea region, but with the construction of canal systems throughout Europe in the 18th and 19th centuries, they soon became well dispersed, arriving in Great Britain by 1824. They were first discovered in the North American Great Lakes in 1988, and are believed to have arrived, like many other invasive species, in the ballast waters of cargo ships. By 1990 they were well established throughout the Great Lakes, and have now moved into the Mississippi River basin. A related species, the Quagga mussel (*Dreissena bugensis*) also invaded the Great Lakes at around the same time and has created additional problems by colonizing soft sediments and deeper water than the Zebra mussels.

A single Zebra mussel can produce up to one million eggs over the course of a spawning season, and these hatch into readily dispersed larvae ('veligers'), that are free-swimming for up to a month. The adults can achieve densities up to hundreds of thousands per square metre, and their prolific growth within water pipes has been a serious problem for the cooling systems of nuclear and thermal power stations, and for the intake pipes of drinking water plants. A single Zebra mussel can filter a litre a day, and they have the capacity to completely strip the water of bacteria and protists. In Lake Erie, the water clarity doubled and diatoms declined by 80–90 per cent soon after the invasion of Zebra mussels, with a concomitant decline in zooplankton, and potential impacts on planktivorous fish. The invasion of this species can shift a lake from dominance of the pelagic to the benthic food web, but at the expense of native unionid clams on the bottom that can become smothered in Zebra mussels. Their efficient filtering capacity may also cause a regime shift in primary producers, from turbid waters with high concentrations of phytoplankton to a clearer lake ecosystem state in which benthic water plants dominate.

The problem of invasive species is now compounded by global climate change. This is weakening the competitive abilities of native plants, animals, and microbes at the upper end of their thermal range, and it is opening new habitat opportunities for species from the warm, temperate, and tropical regions to expand into previously cold, unfavourable lakes. Conservation regions such as national and regional parks are now more important than ever in protecting lake food webs from additional stressors, and in lessening the pressure of invasive species that inevitably accompany land development and the associated expansion of transport routes.

Chapter 6
Extreme lakes

> The composition of lake water that we recognize at the surface, is it maintained at all depths, or instead does it vary, and in what manner?
>
> F. A. Forel

The familiar, upper waters of a lake are often a poor guide to what lies well beneath the surface, and in some lakes the variations with depth are extreme. One of the most striking examples is Lake Vanda in the McMurdo Dry Valleys region of Antarctica, where thick ice overlies the water throughout the year and prevents mixing by the wind. When the first scientists drilled a hole in the ice and lowered their thermistor probe down through the underlying water column, they were surprised to discover that temperatures rose with increasing depth, finally reaching 26°C at the bottom. This temperature inversion of warm water underlying cold was possible because of a strong gradient in salt concentration: while the surface waters of Lake Vanda are fresh and derived from pure glacial ice, its bottom waters are up to three times the salinity of seawater. After a period of lively debate, this unexpected warmth was finally explained as the cumulative effect of sunshine: year after year, century after century, solar radiation in summer has passed through the clear ice and freshwater, and has gradually warmed the dense, salty bottom layer to the unlikely temperatures observed today.

Extreme lakes are waterbodies that have unusual physical, chemical, and biological features, and they are of great scientific interest. Salt water lakes occur in many parts of the world and are often highly productive, with a simplified food chain that supports large flocks of resident or migrating birds. Polar and alpine lakes are strongly affected by snow and ice, and are therefore sensitive to small changes in temperature across the freezing–melting threshold of water. These high latitude and high altitude ecosystems are global sentinels of climate change in the past and present, as well as models for wider understanding of lake microbiology and biogeochemistry. Other extremes among the world's lakes include acid and alkaline waters, geothermal hot water lakes, and waterbodies that periodically erupt, disgorging their liquid and gaseous contents, and creating danger to anyone in the vicinity.

In the most severe lake conditions, only the hardiest of 'extreme-loving' microbes can survive and grow. These 'extremophiles' include halophiles that prefer the highest salinities, psychrophiles (from the Greek 'psukhrós' meaning cold or frozen) that are adapted to perennially cold water, and acidophiles that grow best under low pH. Biochemical and genomic research on these microbes is providing insights into the origins, evolution, and limits of life on Earth, and has yielded unique biomolecules of medical and biotechnological application.

Salt water lakes

One of the many distinguishing features of H_2O is its unusually high dielectric constant, meaning that it is a strongly polar solvent with positive and negative charges that can stabilize ions brought into solution. This dielectric property results from the asymmetrical electron cloud over the molecule, described in Chapter 3, and it gives liquid water the ability to leach minerals from rocks and soils as it passes through the ground, and to maintain these salts in solution, even at high concentrations.

Collectively, these dissolved minerals produce the salinity of the water, measured in terms of grams of salt or dissolved solids per litre. Since 1 litre (L) of water weighs 1,000g, salinity can be expressed as grams per thousand grams or ppt. Sea water is around 35ppt, and its salinity is mainly due to the positively charged ions sodium (Na^+), potassium (K^+), magnesium (Mg^{2+}), and calcium (Ca^{2+}), and the negatively charged ions chloride (Cl^-), sulfate (SO_4^{2-}), and carbonate (CO_3^{2-}).

These solutes, collectively called the 'major ions', conduct electrons, and therefore a simple way to track salinity is to measure the electrical conductance of the water between two electrodes set a known distance apart. Lake and ocean scientists now routinely take profiles of salinity and temperature with a CTD: a submersible instrument that records conductance, temperature, and depth many times per second as it is lowered on a rope or wire down the water column. Conductance is measured in Siemens (or microSiemens (μS), given the low salt concentrations in freshwater lakes), and adjusted to a standard temperature of 25°C to give specific conductivity in $\mu S/cm$.

All freshwater lakes contain dissolved minerals, with specific conductivities in the range 50–500$\mu S/cm$, while salt water lakes have values that can exceed sea water (about 50,000$\mu S/cm$), and are the habitats for extreme microbes, such as the halophilic, green algal flagellate *Dunaliella* and salt tolerant archaeons (Haloarchaea) that have biochemical strategies to contend with such a high level of salinity stress. Deep Lake in the Vestfold Hills of Antarctica is so saline (270ppt) that the water never freezes and one could row out into the middle of the lake in mid-winter while the surrounding landscape is deeply frozen; it would be best to stay out of the water, however, since the temperature of the liquid brine is around −18°C.

Saline waters account for a vast total area, and hold a number of records among the world's lakes. The largest lake in the world is the Caspian Sea, extending over 371,000km^2 with a maximum depth of 1,025m. It has a moderately high salinity (12ppt) that is derived from terrestrial rather than marine sources; in contrast, the Black Sea exchanges water with the Mediterranean Ocean and is therefore considered a marine system, not a lake. Like many saline lakes, the Caspian Sea is an ancient waterbody that occupies a tectonic basin, with numerous endemic species including a landlocked seal, *Pusa caspica*. One of the world's oldest lakes is the moderately saline (6ppt) Lake Issyk-Kul (meaning 'warm lake') that lies in the Tian Shan Mountain area of Kyrgyzstan. This large, deep lake (6,300km^2; maximum depth 702m) has an age rivalling that of Lake Baikal (around twenty-five million years) and supports a diverse fauna including endemic species. The Dead Sea at 400m below sea level is the lowest lake in the world and one of the saltiest: 342ppt, about ten times that of sea water. Salt lakes also occur at extreme high altitudes, including on the Tibetan Plateau and the altiplano of Bolivia and Peru. Despite all of these unusual features, saline lakes have been considered expendable because of their often remote location and salty, undrinkable water. However, their high value for migratory birds and importance to rare species has placed them on the front line of conservation battles in several parts of the world.

Conflicting views about the value of saline lakes were especially apparent in the long and ultimately successful battle to save Mono Lake, California. When Mark Twain visited the region in the early 1860s, he called the lake a 'solemn silent, sail-less sea' that lay in a 'hideous desert'. Yet like many saline waterbodies, Mono Lake is a place of stunning beauty, bountiful plankton, and immense flocks of water birds. The lake is considered a 'triple water' in that its saltiness is due to three components: carbonate (hence it is considered a soda lake), chloride, and sulfate. If you dip your hand into the water, it has a slippery, soapy feel to it; and then as the water dries rapidly in the desert sun, your hand will be left covered with a film of salt, like a thin, white glove.

28. Tufa towers at Mono Lake, California.

Underwater springs flow into Mono Lake, and when these cold freshwaters containing calcium meet the saline lake water, calcium carbonate precipitates out as the mineral calcite to produce pillars called tufa towers. This process is also aided by cyanobacterial films that coat the tufa towers, and their removal of CO_2 by photosynthesis shifts the equilibrium towards more carbonate precipitation. Many of these impressive towers are exposed around the edge of the lake by ancient and modern falling lake levels, and can be several metres in height (Figure 28).

Mono Lake lies on the eastern side of the Sierra Nevada mountain range of California, at the edge of a vast, high, desert region called the Great Basin. This area once contained extensive freshwater lakes, but these ancient waters have long since evaporated to leave behind salt pans and salt lakes. The largest of these residual waterbodies is Great Salt Lake, Utah (4,400km^2; maximum depth 14m), with 50–270ppt of salt, depending on its fluctuating water level. Great Salt Lake, Mono Lake, the Caspian Sea, and many other saline waterbodies, are 'endorheic', meaning they have no outflow. The large evaporative losses of Mono Lake water were offset by freshwater streams that recharged the lake each year with snowmelt from the Sierra Nevada Mountains. But water planners for the city of Los Angeles were casting their efforts far and wide to meet the needs of the rapidly growing population, and began to divert those Sierra Nevada meltwaters and channel them 560km through aqueducts to the city. The first major diversion began in 1941, and from that point onwards Mono Lake began to shrink in size and become increasingly saline. From the 1940s to the 1970s, the lake level fell by around 15m, and the salinity doubled from 40 to 80ppt. Water usage by the city of Los Angeles had already caused the complete drying up of Owens Lake, another endorheic lake in the region, and it seemed that Mono Lake was on a similar path to extinction.

The fortunes of Mono Lake began to reverse when a group of undergraduates from the University of California Davis and Stanford University teamed up to undertake a summer research programme on the ecology the lake in 1976. Their study drew attention to the highly productive food web of this lake, based on two hardy invertebrates: the alkali fly (*Ephydra hians*) that undergoes its larval and pupal stages in the water at the edge of the lake, in the past harvested as food by the Kuzedika Native Americans, and the brine shrimp (*Artemia monica*) that achieves a population of trillions each year in the lake. The brine shrimp feeds on a minute (less than 3μm) salt-tolerant green alga called *Picocystis*.

The student research at Mono Lake showed that the flies and shrimp were the food resources for huge numbers of migratory bird species that used the lake as an important stopover on their flyways each summer, including some 50,000 gulls, 80,000 phalaropes, more than a million eared grebes, and many other species. Most importantly, they showed that rising salinities would push the brine shrimp to extinction. This collapse of food would be compounded by the connection of islands to the shore because of the falling lake levels, which would expose nesting birds such as the California gull to coyotes and other predators. Members of this group, led by ornithologist David A. Gaines, went on to found the Mono Lake Committee that took the city of Los Angeles to court, while raising public awareness of the ecological value of the Mono Lake ecosystem and its dire trajectory. The Committee's tireless efforts over a marathon fifteen years of court battles eventually won legal and political support, culminating in the full re-instatement of inflows and a rise in lake levels. The lake is now a unique conservation park that draws many visitors each year as well as the prolific flocks of migrant water birds.

Polar and alpine lakes

Lakes at high latitudes and altitudes encompass a diverse range of ecosystem types, from the numerous floodplain lakes that wax and wane on Arctic river deltas, to the vast deep waters of Great Bear Lake in northern Canada (31,153km^2; maximum depth 446m), highly stratified waterbodies such as Lake Vanda, Antarctica, and deep, clear alpine lakes such as Lake Redon (2,240m above sea level; maximum depth 73m) in the Pyrenees, first studied by the eminent Catalan ecologist Ramon Margalef. Despite this variety of habitat types, polar and alpine lakes hold a number of features in common, including their remoteness from cities and the direct effects of pollution. This has made such lakes ideal sites to track the long range dispersal of heavy metals and organic contaminants. For example, at Lake Redon, organic contaminants such as DDT and PCBs have been detected in the water many decades after these compounds were banned from use in Europe, indicating their arrival from thousands of kilometres away, and the need to collaborate globally in the control of poisons in the biosphere. Other contaminants have been detected that arrive from regional sources (e.g. hexachlorocyclohexane pesticides that are used in agriculture in southern Europe). The remoteness of polar and alpine lakes has also been of great interest for biogeographical studies, and while there is some evidence of the cosmopolitan distribution of certain cold-tolerant microbes, for example some freshwater cyanobacteria, other studies of these isolated, island-like ecosystems have shown that there are regional assemblages, including cyanobacteria and single-cell eukaryotes (protists).

Another common feature of polar and alpine lakes is their intimate association with the cryosphere, the ensemble of snow- and ice-containing environments throughout the world. These lakes are covered by thick ice for much or even all of the year. This ice cap, often overlaid by snow, restricts the light available for primary production, and this effect is compounded in the polar regions by up to three months of continuous winter darkness. A major impact of climate warming on such lakes is the extension of the ice-free season, with earlier ice melt in spring and later freeze-up at the end of summer. This results in not only more light for photosynthesis, but also more upward mixing of nutrients for algal growth in the surface waters. However, under these open water conditions, the lake biota also has to contend with greater exposure to potentially harmful ultraviolet radiation, which may influence productivity and species composition.

One group of organisms that is pre-eminently successful in polar and alpine lakes is the 'psychrotolerant' cyanobacteria that can withstand extreme cold and even complete freeze-up, but grow more rapidly at warmer temperatures (hence they are not 'psychrophiles' that prefer the cold). These micro-organisms bind together sand and silt particles with their filaments and a mucilaginous glue made of sugary compounds to produce a thick biofilm or 'microbial mat' that coats the bottom of lakes, ponds, and streams. The mats are often bright pink or orange due to carotenoids that prevent UV-B damage from the bright incident sunshine, and they range in thickness from a fraction of a millimetre to several tens of centimetres. The most spectacular communities have been found at the bottom of permanently ice-covered Antarctic lakes, such as Lake Untersee, where they produce dome-like structures that bear a close resemblance to the earliest fossils ('stromatolites') on Earth.

In cold, ice-covered waters, cyanobacterial mats and films often occur in association with mosses, and have high concentrations of red and blue proteins called 'phycobiliproteins', which capture light for photosynthesis with high efficiency. These communities can dominate the primary production and biomass of polar and alpine lakes. Analyses of their microbiome composition is showing that although the dominants are cyanobacteria there are hundreds if not thousands of other microbes present: other bacteria as well as archaea, viruses, and eukaryotes such as diatoms and small invertebrate animals that make their home in nutrient-rich microhabitats provided by the cyanobacterial mat. It has been suggested that eukaryotic cells (protists) might have found refuge and thrived in these biofilms coating the base of melt pools on ice (as they do today on polar ice shelves and glaciers) during the 'Snowball Earth' glaciations that extended over most of the globe between about 720 and 635 million years ago.

Polar and alpine lakes are also useful model systems to study and better understand how elemental cycles function in the aquatic environment, and how lakes are affected by inputs from their surroundings. Some of the best natural laboratories for such studies are the permanently layered lakes in the polar regions, such as lakes Vanda, Fryxell, Hoare, Bonney, Joyce, and Miers in the McMurdo Dry Valleys region of Antarctica. These waterbodies are called 'meromictic lakes', meaning only partially mixed. Another lake district containing these saline, permanently layered waters occurs at the opposite pole, in the Canadian High Arctic. These lakes were first discovered by a military research expedition in 1969, and with tactical formality were named lakes A, B, C.... The lakes still retain these unexciting names from the Cold War era, but the letters belie their many unusual and intriguing properties.

Lake A (latitude 83°N; maximum depth 128m) lies in the national park called 'Quttinirpaaq', an Inuit word meaning 'Top of the World', and occupies a valley that some 5,000 years ago was filled with seawater as a fjord connected to the Arctic Ocean. With the melting of the Arctic ice sheets and the associated loss of massive pressure from above, the valley rose up out of the sea, cutting off the fjord as an isolated lagoon of Arctic Ocean water. Melting snow and glaciers produced dilute waters that discharged into the lake and floated on top of the dense, salty water beneath. The resultant layered nature of Lake A today can be seen by its salinity profile (Figure 29): low conductivity meltwaters occur beneath the ice as a surface layer, while at about 11m there is a sudden rise in salinity that continues down to the ancient seawater that was trapped in the valley thousands of years ago.

The salinity profile of Lake A provides clues to its geological history, while the temperature profile provides a record of more recent change. Its thermal profile shown in Figure 29 indicates some summer warming under the ice, likely associated with the

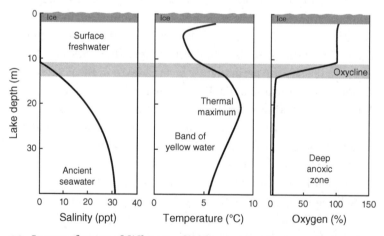

29. **Layers of water of different salinities, temperatures, and oxygen in ice-capped Lake A in the Canadian High Arctic.**

warm, low-salinity inflows, followed by a decrease in temperature in the deeper, underlying water. But then there is an unexpected rise to the thermal maximum at about 22m depth. As in Lake Vanda, Antarctica, the reason for this rise is the gradual warming by sunshine penetrating through the ice and water to this depth.

The surface waters of Lake A are fully saturated in oxygen, recharged in summer by ephemeral streams that flow from melting snowbanks, but deeper in the lake the oxygen values plunge across the oxygen gradient ('oxycline') to below the limit of detection, and these anoxic conditions extend to the bottom of the lake. The colour and smell of the water changes dramatically with depth and is further evidence of this layering. For example, a band of yellow-coloured water occurs at 28–30m depth, and is the result of green photosynthetic sulfur bacteria that capture light for energy and use H_2S (instead of H_2O used by plants) to reduce CO_2 to sugars, in the process producing yellow particles of elemental sulfur. The rotten-egg smell of H_2S is obvious in water samples brought up from 30m and below.

Molecular tools based on nucleic acids (DNA and RNA) offer a powerful set of approaches towards analysis of the layering of different microbial communities, and are now used routinely in lake studies. These are providing insights into microbial diversity and biogeochemical processes that were previously intractable because most members of the aquatic microbiome have not been brought into culture and cannot be distinguished under a microscope. Once DNA has been extracted and its nucleotides (the A, G, C, T alphabet of nucleic acids) sequenced, the genetic relationships can be explored by producing a tree-of-life diagram, where the distance between samples or species is a measure of relatedness. One of the strengths of this approach is that all data are shared in an international database (GenBank), which provides a huge (200 million records at present), ever improving reference source for these sequence comparisons.

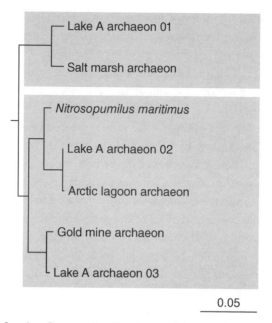

30. Tree showing the genetic relatedness of three archaeon strains in Lake A, and their close relationship to archaea from certain other habitats.

An example of this molecular approach applied to Lake A is shown in Figure 30. First, water was sampled from 10–12m depth where there was a sharp drop in oxygen (Figure 29). This is always an interesting place to look for new microbes because oxygen gradients usually contain a blended cocktail of oxidized and reduced chemicals, favouring a variety of microbial lifestyles. The DNA sequences of ribosomal genes, which have been found to be especially good for differentiating species, were then compared. The DNA results from Lake A showed that the three lake microbes all clustered in the archaea part of the tree (the 0.05 scale bar in Figure 30 indicates a DNA sequence difference of 5 per cent), and that they were in close proximity to the species *Nitrosopumilus maritimus*, an archaeon that oxidizes ammonium to nitrite. The chemical habitat at 10–12m in Lake A (Figure 29) would be ideal for energy production based on ammonium oxidation, with oxygen diffusing down into the oxycline from above and ammonium diffusing up from the anoxic waters below.

DNA-based methods are proving to be valuable in exploring the most extreme of all polar aquatic habitats: subglacial lakes. These waters remain liquid throughout the year, but lie hundreds or thousands of metres beneath the Antarctic ice cap. The first of these was unexpectedly discovered under the research station Vostok, set up by the Soviet Union near the south geomagnetic pole during the International Geophysical Year (1957/8). Radio-echo sounding showed that the 3,750m-thick ice sheet at this location was underlain with liquid water to an astonishing 1,000m depth. Continued geophysical measurements revealed that this hidden water body, named Lake Vostok, is of vast extent, with an area of 14,000km^2 and an estimated volume of 5,400km^3. This quantity of water greatly exceeds many other large lakes of the world, such as Lake Ontario (1,640km^3). The discovery of an aquatic environment sequestered beneath the ice raised questions of great scientific as well as public interest: is Lake Vostok a tectonic basin of sterile water? Or could it be an active lake ecosystem that has somehow functioned for millions of years in the absence of life-giving sunlight?

The quest to look for life in Lake Vostok was especially motivated by astrobiologists, scientists who are interested in the origin, evolution, and limits of life on Earth, and in the conditions that might allow life to exist in places beyond Earth. Liquid water has been detected elsewhere in the Solar System, for example beneath the thick ice crusts of Europa, the smallest of the moons orbiting Jupiter, and Enceladus, the sixth largest moon of Saturn. Lake Vostok seemed like an appropriate analogue in considering the prospect of ecosystems in such places, and to develop the sterile, ice-penetrating technologies that would be needed to retrieve chemical and potentially biological samples from these environments. An even more compelling motivation came from the ongoing surveys of the Antarctic ice sheets, with the discovery that there are hundreds of subglacial lakes (mostly much smaller than Lake Vostok), and that many of these are connected in vast Amazonian-size basins of flowing waters, hidden beneath the thick ice. This would make subglacial waters one of the world's great ecosystem types, with a potentially large, downstream influence on the coastal Southern Ocean, into which these waters ultimately discharge. Geophysicists are also greatly interested in these subglacial liquid environments given that ice sheet stability and flow will also be affected by the presence of lubricating water at the ice–land interface, with effects on global climate, ocean circulation, and sea level.

The first attempts to sample these subglacial lakes were marred by setbacks and uncertainty. The Russians had already drilled to great depth at Vostok Station to obtain a record of past climate change. Their 3.4km-long ice core provided an unprecedented view of the natural cycles of greenhouse gas concentrations over the last 400,000 years, showing past maxima that we have now vastly exceeded through human activities. However, kerosene aircraft fuel had been used to keep the hole open over the ten years of drilling, and when the drill finally broke through the ice into the lake in February 2012, the resultant lake samples were possibly contaminated with this fluid, making it difficult to resolve any native microbial community. In December 2012, a British research team attempted to sample Lake Ellsworth, a 150m-deep subglacial waterbody in west Antarctica overlain by 3,400m of ice. They used a sterile, hot-water drilling system to ensure no contamination of the subglacial waters, but unfortunately the fuel supply ran out before drilling could progress beyond 300m depth into the ice.

In January 2013, an American team broke through into Lake Whillans in west Antarctica using hot-water drilling and a series of protocols to eliminate microbial and chemical contaminants. This lake is known to be an active system that fills and drains regularly, as indicated by shifts in its surface ice elevation, and at the time of sampling it was 2.2m deep and overlaid by 800m of ice. The team applied DNA sequencing methods to determine the microbial community structure and they found a diverse microbiome in the water, with a prevalence of ammonium-oxidizing archaea as in the oxycline of Lake A, and bacteria similar to nitrite oxidizers that have been isolated from Arctic permafrost. A sediment core was also taken from the lake and its microbial community included methane-consuming bacteria at the surface, and methane-producing archaea at depth.

Many questions remain, such as whether there is a microbial network of biological interactions with eukaryotic cells and viruses, and how representative is the Lake Whillans microbiome. However, these first results provided compelling evidence that the subglacial environment is a vast living ecosystem based on microbes that use inorganic chemicals for energy, along with other microbes that derive their energy from organic materials. Over the decades ahead, the exploration of Antarctic subglacial lakes will continue to be an exciting frontier for 'extreme limnology', and will also provide insights into how life survived beneath the vast ice sheets that covered much of the world during glacial epochs in the past.

Exploding lakes

At the opposite extreme to the cold water ecosystems of polar and alpine regions, geothermal waters also hold great interest for lake scientists and microbiologists. Here once again, life has been pushed to its limits of survival, yet a surprising variety of microbial extremophiles has the ability to thrive under these harsh conditions of low pH and scalding hot temperatures. The genomic techniques described above have been applied with great success to these waters, and have revealed unusual species and strategies for survival in these severe habitats. Some of the biomolecules found in the microbes living in these lakes have also proved to be of immense commercial value, and bioprospection of microbes from geothermal ecosystems has resulted in the development of novel products for use in biotechnology and in the biomedical industry. One of the most well known is the enzyme called 'Taq polymerase'. Taq is the abbreviation for *Thermus aquaticus*, a microbe first isolated from a hot pool at Yellowstone National Park, USA, and the original source of this enzyme that is used in an important technique for amplifying DNA for analysis, known as the polymerase chain reaction (PCR). *Thermus aquaticus* grows naturally at temperatures from 50 to 80°C, and its heat-stable DNA polymerase therefore proved ideal for the alternating high temperatures used in PCR.

A perennial hazard of living in an active geothermal region is that the ground and its associated waters have a tendency to blow up from time to time. This may be in the form of a hydrothermal explosion crater, where trapped gases including water vapour finally exceed the pressure resistance of their surrounding rock and soil, explode out of the ground, and leave behind a large hole that then fills with water to form a lake. The craters of active volcanoes may also fill with water that can be ejected from the lake during eruption, or drain through a breached wall of volcanic ash.

One such example is the crater lake on Mount Ruapehu, shown in Figure 31. The waters of this lake fluctuate greatly in temperature, with values up to 60°C, and a highly acidic pH that can be as low as 0.9. The volcano has experienced three major eruptions over the last 150 years, with smaller eruptions at more frequent intervals. The mountain is now carefully monitored and is installed with a seismic alarm system on its snowy slopes, which in case of eruption warns skiers to move rapidly to safety away from the slurry of lake water and sediments ('lahar') that could begin flowing down the valleys. This monitoring was prompted by the tragedy of 23 December 1953, when, following an eruption, Ruapehu crater lake burst its retaining dam of ash.

31. Highly acidic lake in the active volcanic crater of Mount Ruapehu, New Zealand.

The lahar flowed down a river gorge, collapsing a railway bridge on the main train line. Unaware of this sudden event that had happened only minutes earlier, the night express train and its first six carriages plunged into the gorge, killing 151 of the passengers onboard.

Volcanic lakes can pose other threats to human life, including as the result of their supersaturation of gases. Lake Nyos in Cameroon occupies the crater of an extinct volcano, but a magma chamber beneath the lake leaks carbon dioxide into the water, resulting in extreme concentrations that can be suddenly released at the surface due to landslides or earthquakes. A massive cloud of carbon dioxide was emitted from the lake in 1986, and suffocated 1,746 people and 3,500 livestock in the surrounding area. Since that time, tubes have been installed into the lake to vent the gases from its deep waters and to lower the risk of explosion. A similar accumulation of gas is known in Lake Monoun, also in Cameroon, where an eruption in 1984 released large quantities of carbon dioxide that suffocated and killed thirty-seven people.

A third, much larger, lake that is charged with volcanic gases is Lake Kivu, on the border of Rwanda and the Democratic Republic of Congo. The bottom waters of this large deep lake (2,700km^2; maximum depth 480m) interact with a volcano, which has resulted in the build-up of methane as well as carbon dioxide. These gases emanate from time to time from the lake, and pockets of toxic, CO_2-enriched air are locally called 'mazuku', a Swahili word meaning 'evil winds'. A silver lining to this dangerous situation is that the methane at depth is also a potential source of fuel for power generation. A plant has now been installed at the lake that pumps up water, and extracts and burns the methane. This generates around 26 megawatts (MW) of electricity, while also reducing the gas content and risk of catastrophic explosion of the deep waters of the lake.

Chapter 7
Lakes and us

> Humans exert a more powerful effect than any other animal on Nature and its inhabitants.
>
> F. A. Forel

When François Forel began his catalogue of the plants and animals of Lake Geneva, the first species he placed on the list was *Homo sapiens*. He introduced the notion that humans are not only part of the lake ecosystem through activities ranging from lake shore development to transport of goods and people (Figure 32), but also have the capacity to do great damage to a lake and its ability to provide essential services such as fisheries and safe drinking water. He observed that lake levels change through human intervention as well as natural causes, and he was an expert witness in litigation against the city of Geneva and its defective control structure at the outflow of the lake. Little did he realize the magnitude of dam-building and proliferation of artificial lakes that would take hold of human society in the 20th century and that continues with fervour today in developing countries.

Forel surveyed the vast expanse of Lake Geneva (area of $580km^2$ and volume of $89km^3$) with the sense that it would forever remain a limitless source of high-quality drinking water for all residents of the lake. Yet later in the 20th century, this lake, like so many

32. A traditional merchant vessel on Lake Geneva in the 19th century.

others around the world, began to experience the effects of eutrophication, with a rapid decline in water quality, depletion of its bottom water oxygen, and proliferation of algae. For Lake Geneva and all freshwater resources, the greatest challenges may lie ahead with global climate change and the associated impacts of increased temperatures, shifts in mixing patterns, extreme weather events, changing water supply, and altered habitat conditions for native and invasive species.

Dams large and small

For thousands of years, humankind has dammed and impounded waters to create artificial lakes and ponds. Up until the late 19th century, almost all of these were small in scale and included structures for crop irrigation, livestock watering, flood control, and domestic water supply, waterbodies for aesthetic and cultural purposes, millponds for water power, and impoundments for fish farming. Over the course of the 20th century, large-scale projects for navigation and hydroelectricity became symbols of progress, and brought considerable economic benefits along with a vast expansion of lake waters. The reservoirs of Europe currently total

100,000km² in area, including behind two large dams on the Volga River, the Kuybyshevskoye (6,450km²) and Rybinskoye (4,450km²) reservoirs. The World Register of Dams currently lists 58,519 'large dams', defined as those with a dam wall of 15m or higher; these collectively store 16,120km³ of water, equivalent to 213 years of flow of Niagara Falls on the USA–Canada border. One of the largest hydroelectric schemes in the world is the James Bay complex in northern Quebec, Canada, which began operation during the late 1980s; this covers a total reservoir area of 11,800km² and has a current generating capacity of 16,500MW, with further expansions in progress.

Although dam-building has slowed or even reversed in the Western world, there is a new boom phase underway in Asia, Africa, and South America. The Three Gorges dam on the Yangtze River (1,084km²; dam height 181m) in China began operation in 2012 and is now the largest hydroelectric power station in the world in terms of capacity (22,500MW). Around a hundred large dam projects are in advanced planning or construction in Africa, including the 145m-tall Grand Ethiopian Renaissance Dam on the Blue Nile River. More than 300 dams are planned or under construction in the Amazon Basin of South America, including the Belo Monte dam complex on the Xingu River.

Reservoirs have a number of distinguishing features relative to natural lakes. First, the shape ('morphometry') of their basins is rarely circular or oval, but instead is often dendritic, with a tree-like main stem and branches ramifying out into the submerged river valleys. Second, reservoirs typically have a high catchment area to lake area ratio, again reflecting their riverine origins. For natural lakes, this ratio is relatively low; for example this ratio for Windermere and Wastwater in the English Lake District is around 16; for Lake Geneva it is 13.8, while for Lake Tahoe the ratio is only 2.6, and it is a factor contributing to the long water residence time of this lake (650 years). In contrast, for Lake St-Charles, the dammed drinking water reservoir for Quebec City, the catchment to lake area ratio is 46; for Lake Mead on the Colorado River behind the Hoover Dam, USA, it is 640; and for the Three Gorges reservoir this ratio is 923. These proportionately large catchments mean that reservoirs have short water residence times, and water quality is much better than might be the case in the absence of this rapid flushing. Nonetheless, noxious algal blooms can develop and accumulate in isolated bays and side-arms, and downstream next to the dam itself.

Reservoirs typically experience water level fluctuations that are much larger and more rapid than in natural lakes, and this limits the development of littoral plants and animals. Another distinguishing feature of reservoirs is that they often show a longitudinal gradient of conditions. Upstream, the river section contains water that is flowing, turbulent, and well mixed; this then passes through a transition zone into the lake section up to the dam, which is often the deepest part of the lake and may be stratified and clearer due to decantation of land-derived particles. In some reservoirs, the water outflow is situated near the base of the dam within the hypolimnion, and this reduces the extent of oxygen depletion and nutrient build-up, while also providing cool water for fish and other animal communities below the dam. There is increasing attention being given to careful regulation of the timing and magnitude of dam outflows to maintain these downstream ecosystems.

Dams create new lakes for hydropower, irrigation, and drinking water, but the environmental costs are not always apparent at the time. Lake Urmia, a great salt lake (5,200km^2 at its maximum extent) in Iran that is renowned for its bird life has shrunk to 10 per cent of its original size, in part because its three primary inflows have been dammed for irrigation and hydropower. The resultant deposits of salt are blown around in the wind and affect farmlands as well as human health. Similar environmental problems are also encountered at the vast Aral Sea (Kazakhstan/Uzbekistan), which shrank from 68,000km^2 in the 1960s to around 7,000km^2

by 2005 as a result of inflow diversions for irrigation. A dam has now been built at the northern corner of this lake to retain water, dilute the salt, and restore the fishery in a small part of the original basin.

Dam-building can have wide ranging consequences for the original residents of the river basin, both human and animal. The Belo Monte project will flood and perturb lands used by thousands of Amazonian Indians, and the cultural impacts have generated international concern and protest. To build the Three Gorges Reservoir, some 1.2 million people were displaced, including the entire populations of thirteen cities. The dam is now an impediment to animal migration, including the Yangtze sturgeon and other endangered fish species, but the greatest effects may be downstream, with increased ship traffic and shifts in the annual flood regime. There is evidence that lower water levels in the Yangtze River floodplain due to reservoir operations are conducive to the transmission of parasitic flatworms from aquatic snails to humans, causing the severe disease 'snail fever' or schistosomiasis; a rising incidence of this disease has also accompanied other hydro-developments such as Egypt's Aswan dam. The lower water levels are putting wetlands at risk, and reducing the connections among habitats for fish and other aquatic animals. Perturbation of the flow regime may also affect the spawning, egg hatching, growth, and migration activities of native species that have evolved in response to the natural cycle of water fluctuation. The impacts of dams on fish biodiversity are of special concern in tropical river basins of the Amazon, Congo, and Mekong rivers, which currently hold around 4,200 species, of which 60 per cent are endemic. In total for these three basins, some 840 dams are currently operating or under construction, while another 445 are in various stages of planning.

The downstream effects of dams continue out into the sea, with the retention of sediments and nutrients in the reservoir leaving less available for export to marine food webs. This reduction can also lead to changes in shorelines, with a retreat of the coastal delta and intrusion of seawater because natural erosion processes can no longer be offset by resupply of sediments from upstream. Severe erosion has already been observed in the Yangtze River delta since the Three Gorges Dam began to operate. An additional effect of reservoirs is the mobilization of mercury from flooded vegetation and soils. This is methylated by bacteria to the more toxic form, methyl mercury, which can be further concentrated at each step of the food chain and transferred to downstream marine waters.

Many of us throughout the world depend on reservoirs for flood control, water supply, electricity, and economic well-being, and the ecosystem services provided by dammed lakes are now an integral part of our civilization. Dam-building continues in much of the world, and there are calls for an expansion of construction efforts to mitigate the effects of climate change on future water availability, reduce reliance upon fossil fuels, and keep up with the ever increasing needs of our global population, estimated to increase by another three billion over the course of this century. There are salutary reminders from the past, however, that the costs of such projects can often be underestimated, the benefits oversold, and the human and environmental consequences given inadequate thought and consideration, to the long-term detriment of social and ecological values.

Greening of the world's freshwaters

One of the most serious threats facing lakes throughout the world is the proliferation of algae and water plants caused by eutrophication, the overfertilization of waters with nutrients from human activities. This issue came to the fore in the mid-20th century with the realization that while lakes become gradually more nutrient-rich with time, lose their clarity, and eventually infill with sediment and plant growth, this slow, natural process can be hugely accelerated by increasing nutrient inputs from human activities in the surrounding catchments. The resultant nutrient-rich 'eutrophic' or (with even more enrichment) 'hypertrophic' waters are commonly referred to in the popular press as 'dead lakes'. The toxic algae and lack of oxygen in such lakes can result in death and extinction, however the term is a misnomer because eutrophic waters teem with aquatic life, but unfortunately dominated by noxious species that severely impair fishing, drinking water usage, and other ecosystem services.

Nutrient enrichment occurs both from 'point sources' of effluent discharged via pipes into the receiving waters, and 'nonpoint sources' such the runoff from roads and parking areas, agricultural lands, septic tank drainage fields, and terrain cleared of its nutrient- and water-absorbing vegetation. By the 1970s, even many of the world's larger lakes had begun to show worrying signs of deterioration from these sources of increasing enrichment. In Lake Geneva, for example, the winter Secchi depth plunged from the values measured by Forel of 15–20m in the 1870s, to at best 10m in the 1970s. Forel reported that the bottom waters of Lake Geneva were well oxygenated, even at 300m towards the end of stratification, but a hundred years later, deep water oxygen concentrations had fallen to hypoxic values of less than 2mg/L that excluded bottom-dwelling animals from certain parts of the lake and likely caused the extinction of some species such as the blind shrimp (Figure 22).

A sharp drop in water clarity is often among the first signs of eutrophication, although in forested areas this effect may be masked for many years by the greater absorption of light by the coloured organic materials that are dissolved within the lake water. A drop in oxygen levels in the bottom waters during stratification is another telltale indicator of eutrophication, with the eventual fall to oxygen-free (anoxic) conditions in these lower strata of the lake. However, the most striking impact with greatest effect on ecosystem services is the production of harmful algal blooms (HABs), specifically by cyanobacteria.

In eutrophic, temperate latitude waters, four genera of bloom-forming cyanobacteria are the usual offenders: *Microcystis*, *Dolichospermum* (formally known as *Anabaena*), *Aphanizomenon*, and *Planktothrix*. These may occur alone or in combination, and although each has its own idiosyncratic size, shape, and lifestyle, they have a number of impressive biological features in common. First and foremost, their cells are typically full of hydrophobic protein cases that exclude water and trap gases. These honeycombs of gas-filled chambers, called 'gas vesicles', reduce the density of the cells, allowing them to float up to the surface where there is light available for growth.

Put a drop of water from an algal bloom under a microscope and it will be immediately apparent that the individual cells are extremely small, and that the bloom itself is composed of billions of cells per litre of lake water. In the example shown in Figure 33, each cell is around 5μm in diameter, with a conspicuous bright spot caused by its gas vesicles that scatter the light. For such a tiny, solitary cell, its floating speed to the surface would be so slow as to be almost useless, but by combining that buoyancy in multicellular colonies, the flotation rate can be impressively fast, up to 5m per hour for *Microcystis* colonies.

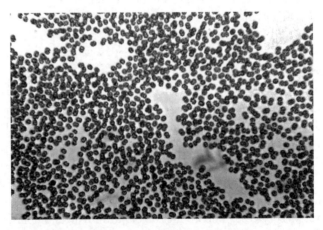

33. Photomicrograph of the toxic bloom-former *Microcystis aeruginosa*.

This flotation ability can also be regulated. During the day, the cells capture sunlight and produce sugars by photosynthesis; this increases their density, eventually to the point where they are heavier than the surrounding water and sink to more nutrient-rich conditions at depth in the water column or at the sediment surface. These sugars are depleted by cellular respiration, and this loss of ballast eventually results in cells becoming less dense than water and floating again towards the surface. This alternation of sinking and floating can result in large fluctuations in surface blooms over the twenty-four-hour cycle.

The accumulation of bloom-forming cyanobacteria at the surface gives rise to surface scums that then can be blown into bays and washed up onto beaches. These dense populations of colonies in the water column, and especially at the surface, can shade out bottom-dwelling water plants, as well as greatly reduce the amount of light for other phytoplankton species. The resultant 'cyanobacterial dominance' and loss of algal species diversity has negative implications for the aquatic food web, especially since these large colonial forms are difficult for the zooplankton to filter and ingest. Additionally, cyanobacteria tend to be deficient in essential fatty acids, and are therefore a poor quality food for animals. This negative impact on the food web may be compounded by the final collapse of the bloom and its decomposition, resulting in a major drawdown of oxygen.

Bloom-forming cyanobacteria are especially troublesome for the management of drinking water supplies. First, there is the overproduction of biomass, which results in a massive load of algal particles that can exceed the filtration capacity of a water treatment plant, especially if its intake is located at a depth or in a bay where these floating colonies accumulate. Second, there is an impact on the taste of the water. Cyanobacteria in general produce a broad spectrum of 'secondary compounds': biologically derived chemicals that do not participate in the primary processes of photosynthesis, respiration, and growth. For many, if not most of these compounds, it is not clear why cyanobacteria bother to produce them, although there is no shortage of hypotheses, ranging from chemical warfare on competing phytoplankton species and toxic dissuasion of herbivores, to the mobilization of trace metals and communication among cells. A number of these biochemicals produce unpleasant tastes and odours, including the musty, earthy odours of geosmin and 2-methyl isoborneol, the grassy taste of cyclocitrals, and sulfurous compounds such as alkyl sulfide that are released during decomposition. The third and most serious impact of cyanobacteria is that some of their secondary compounds are highly toxic.

Toxic lakes

On Saturday 2 August 2014, the mayor of Toledo, Ohio, held an emergency press conference. Mayor D. Michael Collins announced that residents should not drink or boil the tap water, and that all restaurants would be closed until further notice. The environmental chemists at the city's water treatment plant had detected a spike in a cyanobacterial toxin, microcystin-LR, during their routine testing of water quality, with values that exceeded the World Health Organization (WHO) limit of 1ppb. Toledo's water is drawn from Lake Erie, where cyanobacterial blooms occur each year over vast areas and often contain this toxin, sometimes at levels that exceed the WHO limit by a factor of 100. However, microcystin-LR is mostly retained within the cells and it can normally be removed by filtering off the algal particles. The problem at Toledo was that the toxin had made its way through to the post-treatment or 'polished' drinking water. After further tests and safety procedures, the tap water advisory was lifted the following Monday, but the closure of the water supply in a city of half a million people had lasting public impacts in the USA and Canada, and it refocused attention towards the seriousness of eutrophication and toxic water.

The issue of cyanobacterial toxins (cyanotoxins) has also been of great concern in other parts of the world. Lake Taihu ('Great Lake' in Chinese) is the third largest lake in China and is the drinking water supply to ten million people. Although vast in size ($2,338 km^2$), it is shallow (maximum depth 2.6m) and highly eutrophic, and experiences a continuous bloom of *Microcystis aeruginosa* throughout the year. In 2007, residents of the city of Wuxi shifted to bottled water for a month because of the strange tastes and odours in their tap water, and the possibility of imbibing cyanotoxins from the lake. These concerns continue to this day, with efforts to improve monitoring and pollution control for this important water resource.

Microcystins are a class of water soluble toxins produced by many species of bloom-forming cyanobacteria, but most notably by the cosmopolitan species *Microcystis aerginosa*. Chemically they are classified as peptides: each molecule is a series of amino acids linked together with peptide bonds, just like proteins. Unlike many proteins, however, they are not denatured by boiling, possibly because the amino acids are arranged in a stable ring structure (Figure 34). This sturdy configuration also resists the effects of protein degrading enzymes (proteases) that are exuded by bacterial decomposers. Detailed analytical research has shown

34. The toxic peptide microcystin-LR produced by cyanobacteria.

that although the basic ring structure is the same, the side groups can vary greatly, and more than a hundred chemical variants or 'congeners' of microcystins have been detected in *Microcystis*-rich waters, with microcystin-LR the most toxic.

The resistance of microcystins to water treatment protocols belies their biochemical reactivity. Once inside mammals, these toxins move to the liver where they are taken up by cells and interrupt the activity of key enzymes, specifically phosphatases. This ultimately results in liver damage, with associated oxidative stress to the kidney, brain, and reproductive organs. There is also evidence that microcystins have carcinogenic effects by disrupting microtubule assembly and cellular division. Nausea, vomiting, and gastrointestinal illnesses have been linked to drinking water containing microcystins, but the only known human fatalities were reported from a hospital in Caruaru, Brazil, in 1996. More than a hundred renal patients became ill, and seventy of them died during dialysis sessions with water derived from a reservoir with a *Microcystis* bloom; microcystins were detected in the water purification system of the clinic and also in samples of blood and liver of the patients. There are also many reports of sickness and deaths of dogs and farm animals that have drunk water containing toxic cyanobacterial blooms, including *Microcystis*.

When the toxic effects of *Microcystis* first became known in the 1950s, assays with mice showed that the cyanotoxins caused rapid mortality and they were labelled the 'Fast Death Factor', now known to be microcystins. However, an even more potent, faster acting cyanotoxin was isolated in Canada in the 1960s after several herds of cattle had died from drinking bloom-containing waters, and this came to be known as the 'Very Fast Death Factor'. It was ultimately shown to be an alkaloid, named 'anatoxin-a', with potent effects on the nervous system that can cause death within minutes. This cyanotoxin was first isolated from the nitrogen-fixing species *Dolichospermum (Anabaena) flos-aquae* that is commonly found in eutrophic lakes, but it is now known to be produced by several other species and genera of cyanobacteria.

In addition to microcystins and anatoxin-a, bloom-forming cyanobacteria produce a variety of other compounds that are toxic to wildlife, domestic animals, and humans. These include organophosphates, bioactive amino acids, and paralytic shellfish toxins. Some species produce cell wall materials and other compounds that cause skin irritation and dermatitis, and there are many cases of skin allergy reactions by swimmers in bloom-infested waters. Some of these reports, however, may be due to another cause: larval flatworms (schistosomes) that infect freshwater snails and ducks, but that can also burrow into human skin to cause 'swimmer's itch'.

Clearing the water

Can a eutrophic 'dead lake' be restored to its original, near-pristine condition? This important but ambitious goal requires an understanding of the mechanisms and processes that lead to the overproduction of water plants and algae, especially toxic cyanobacteria. In the latter half of the 20th century, when many lakes of the world were experiencing the effects of rapid population growth and increasing discharge of pollutants into their waters, the discussions revolved around three nutrients: carbon, nitrogen, and phosphorus. The North American soap and detergent industry was reluctant to see any mandated changes to their production of phosphate-rich products, and argued that carbon was the element causing eutrophication. Short-term experiments with bottles of lake water enriched with carbon seemed to support this argument, although some of these bioassays gave equivocal and contradictory results.

The most convincing evidence of how nutrients cause eutrophication and which element often plays the greatest role in lakes came from the work of Canadian limnologist David W. Schindler in the

Experimental Lakes Area (ELA) of Canada. The vast granite lands of northern Canada called the Precambrian Shield contain millions of lakes and ponds that have been scratched out from the rock by recent glacial activity, and a small area of this lake-rich landscape in northern Ontario was set aside in 1968 for whole-lake observations and experiments. Schindler's experiment was elegant in its simplicity, and it produced results at the whole ecosystem level rather than in an artificial laboratory environment. He and his team installed a nylon-reinforced vinyl curtain across the middle of an hourglass-shaped lake (number 226 in the ELA inventory) and one side, the southwest basin, was fertilized with carbon (as sucrose that would be rapidly converted to carbon dioxide by the bacteria) and nitrogen (as nitrate). The other side, the northeast basin, was also fertilized with carbon and nitrogen, but additionally with the third element phosphorus (as phosphate), all in ratios approximating those in discharges from sewage treatment plants.

The results were spectacular (Figure 35). The carbon plus nitrogen enriched side of the lake showed little change in algal biomass, as measured by the photosynthetic pigment chlorophyll. This was especially interesting in that Lake 226, like other Canadian Shield lakes, had low natural levels of dissolved inorganic carbon; if carbon were to have an enrichment effect, this would be one of the best places to look for it. In contrast, the lake on the carbon plus nitrogen plus phosphorus side of the curtain soon developed a noxious algal bloom dominated by nitrogen-fixing cyanobacteria. This bloom gave the water a turbid green appearance, and the Secchi depth shrank from around 3m to 1m. Apart from the differences in many of the water quality variables that were measured, the striking visual contrast in the lake between the two sides of the curtain provided compelling evidence to policy makers: phosphorus is the key nutrient limiting bloom development, and efforts to preserve and rehabilitate freshwaters should pay specific attention to controlling the input of phosphorus via point and nonpoint discharges to lakes.

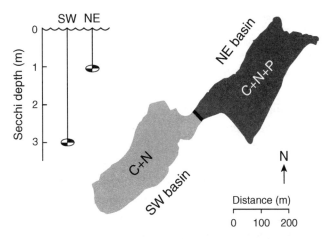

35. **Cyanobacterial bloom development in the CNP-fertilized northeast basin of Lake 226 and the resultant drop in Secchi depth.**

Over the last half century, there has been a focus of attention throughout the world on controlling phosphorus by identifying sources, moving effluent out of the lake basin, installing phosphorus-stripping systems, and regulating the use of phosphorus-rich detergents and other products. One of the earliest examples was Lake Washington, USA, where sewage effluents from the burgeoning city of Seattle were discharged into the lake at ever increasing rates, up to 80 million litres per day in the early 1960s. Work by W. Tommy Edmondson at the University of Washington drew attention to the precipitous decline in water quality of the lake including an increase in nutrients and proliferation of cyanobacteria, and these findings ultimately led to a diversion scheme to redirect the city effluents to the sea. This was implemented in a step-wise fashion over a period of several years, and by 1968 there was no sewage effluent discharged into the lake. Over the five-year period from 1964 to 1969, Edmondson's team showed that there was massive improvement in water quality, with mean algal concentrations in summer declining by a factor of 6, in tandem with a similar decline in winter phosphate concentrations.

Current discussions now focus on the question: is it enough to control only phosphorus? Carbon is in plentiful supply in lakes from the overlying atmosphere and from inorganic and organic sources in the watershed. However, there are multiple reasons for also considering nitrogen. One counter-argument to targeting nitrogen is that nitrogen-fixing cyanobacteria, such as the species that rose to prominence in Lake 226 after carbon + nitrogen + phosphorus enrichment, have an unlimited source of nitrogen that cannot be controlled: gaseous nitrogen in the atmosphere. However, this is not entirely correct in that nitrogen fixers derive only part of their nitrogen requirements from the atmosphere and must otherwise depend on forms such as ammonia, nitrate, and organic nitrogen in the water. Furthermore, one of the most worrisome of bloom-forming cyanobacteria in lakes and reservoirs is *Microcystis*, and this species is unable to fix nitrogen. It produces the nitrogen-rich toxin microcystin (Figure 34; each molecule has ten atoms of N), and there is evidence that the production of this toxin is enhanced by nitrogen enrichment. This toxic species appears to be undergoing a resurgence throughout the world, stimulated by nitrogen- as well as phosphorus-rich fertilizers that spill from agricultural lands out into waterways through increasingly efficient soil drainage systems.

Certain lakes throughout the world lie on catchments that are naturally rich in phosphorus, and are strikingly different in their chemistry from those of the ELA (e.g. lakes on the central volcanic plateau of the North Island, New Zealand, and Lake Titicaca in South America). For these lakes, full control of phosphorus loading is unrealistic. Unbalanced phosphorus control may also lead to macrophyte expansion, since these plants have access through their roots to the bountiful reserves of phosphorus in the sediments, while benefiting from the ongoing nitrogen-enrichment of the overlying water. Finally, freshwater lakes ultimately discharge into the sea, and coastal marine environments tend to be rich in phosphorus, with limiting concentrations of nitrogen. To consider only phosphorus removal may transfer the problem of eutrophication downstream to these receiving waters.

For all of these reasons, the Environmental Protection Agency of the USA and the European Union have recommended control of nitrogen as well as phosphorus removal from effluents. This policy decision has been controversial because nitrogen-removal treatments are expensive and technically more difficult than phosphorus-stripping. The focus upon a single element phosphorus provided a clear, unambiguous target that policy makers and managers could focus upon, and in the process the loading of all nutrients has often been reduced, for example by piping treated effluent out of the basin (e.g. Lake Tahoe and Lake Washington), or by the use of natural and engineered wetlands to remove nitrogen as well as phosphorus. Whatever the local decision, Schindler's results from Lake 226 and his related experiments at the ELA will always be compelling evidence that humans have the capacity to rapidly shift lakes from pristine waters to noxious blooms, and that control of external nutrient supply is essential to preserve our lakes from algal overproduction.

Lake water recovery after nutrient controls have been put in place has not always been as rapid or as complete as was hoped for. Part of the problem is that of 'hysteresis': the trajectory of a lake during its recovery phase after nutrient reductions may follow a different path to the one that led to its degradation, particularly if the lake has undergone a 'regime shift' to persistent noxious blooms, sharply decreased water transparency, and, as a result, loss of underwater plant communities such as algal charophytes (Figure 36). There are many processes that affect this return pathway and that cause a slow-down or inertia to restoration measures. Part of this may be for biological reasons. For example, after many years of cyanobacterial growth, the sediments may contain abundant resting spores and dormant cells that are then an inoculum for ongoing blooms. However, one of the greatest effects causing the slow pace of recovery is nutrient release from

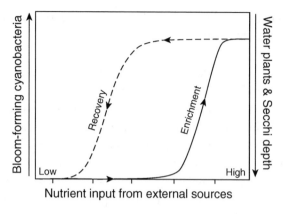

36. **Hysteresis in the recovery versus degradation of a lake.**

the sediments, especially phosphorus, which can accelerate under anoxic conditions (with known exceptions). This is referred to as 'internal loading' to separate it from the external loading on the lake from its surrounding catchment. Such effects are a concern for Lake Erie, for example, where the severe toxic blooms are the result of increases in external loading but are compounded by internal phosphorus release from anoxic sediments (Figure 19). More release means more algal growth, producing more biomass for decomposition and more oxygen depletion towards anoxia. This is a vicious circle that is difficult to stop, and by far the best approach is to protect our lake waters before they reach this oxygen-depleted state.

The future of lakes

In his monograph on Lake Geneva, Forel drew attention to the importance of physical, chemical, biological, and human features of the lake, and the need to incorporate all of these aspects into an integrated synthesis, 'an overview of all the detailed facts, where each specialised study would be supported by the data from other studies'. This type of integrated overview is today at the heart of Earth system science, in which each aspect of the environment, from geophysics to human processes, is considered an interacting part of the global system. An integrated, system-level perspective on lakes is now vitally relevant to managing the world's freshwater resources in the face of rapid climate change at a planetary scale.

Lakes at many sites throughout the world now show evidence of warming, on average at rates that approximate the increases in air temperature. However, there are large variations in the extent of this warming, even among lakes in the same climate region because of differences in depth, wind exposure, and transparency. Extreme weather conditions are likely to accompany climate warming, and in parts of Europe and North America heavy rainfall events have been identified as a factor causing increased inputs of coloured dissolved organic matter and a resultant 'browning' of lakes. This organic enrichment can modify aquatic food webs (Figure 16), and the reduction of transparency means that more solar energy is absorbed in the surface waters, further contributing to warmer temperatures.

Increases in water temperature can have wide ranging effects that propagate throughout the lake ecosystem. Evaporation rates increase with warming, and this may shift the water balance towards net loss, and a reduction in lake levels that is either offset or exacerbated by changes in rainfall. Even small fluctuations in water level can have serious effects on important ecological features. For example, the wetlands around the North American Great Lakes are important for migratory birds and many fish species, and these semi-aquatic habitats are especially vulnerable to minor shifts in water balance. Rising temperatures will also reduce the extent of favourable habitats for cold-water specialists among the flora and fauna, and will facilitate invasion by species from warmer climates.

One of the less obvious effects of climate change in lakes is on the layering (stratification) of the water, with warmer conditions at the surface producing a greater difference in temperature and therefore density between the surface and bottom waters. This stronger stratification provides more resistance to mixing by the wind, and it lessens the exchange of oxygen from the atmosphere to depth in the lake as well as the transfer of nutrients from deep waters to the surface. In Lake Tanganyika, this effect of warming and increased stability appears to have resulted in decreased nutrient supply to the photic zone by mixing, resulting in decreased phytoplankton production and a 30 per cent drop in fish yield. These thermal stratification effects are of special concern for managing noxious blooms of cyanobacteria, which are directly stimulated by warm temperatures and also prefer stable waters for their migration based on gas vesicles and buoyancy control.

Forel described how the majority of residents of the Lake Geneva basin lived in close proximity to the lakeshore and were a dependent part of the lake ecosystem. This latter idea ran counter to the prevailing view at the time and through much of the 20th century that humankind lives on a higher plane than Nature, with full dominion over the land, the air, and the water, and unlimited license to exploit these resources to meet our ever growing needs. At the time that Forel was writing the third volume of his monograph and describing the human history of Lake Geneva, some 56,000 people lived at Lausanne, the world population was around 1.6 billion and atmospheric CO_2 levels were 296ppm; over the subsequent hundred years, these local and global populations both increased by a factor of four, and CO_2 levels rose by 25 per cent. Today, more than 800,000 people draw their water from the lake, and there is ongoing concern about controlling nutrient inputs and contaminants, including emerging pollutants such as pharmaceuticals, microplastics (polyethylene particles less than 5mm in size), and engineered metal nanoparticles (1–100nm) that are increasingly common in the world's freshwaters. Additionally, and like many lakes elsewhere, Lake Geneva has begun to show the effects of climate change, with evidence of warming bottom waters, changes in stratification and mixing, and shifts in the spawning dates of some fish species.

The increasing impacts of population growth and global change on lakes throughout the world are a reminder that while we may be the most powerful entities in the biosphere, we have a close reciprocal relationship with our planetary environment, and a vested interest in protecting its integrity and the ecological services that we crucially depend upon. Lakes are centres of biodiversity, reactive 'slow rivers' that flow, mix, and counter-flow, conduits to the atmosphere and ocean, integrators of the surrounding environment, and sentinels of change in the past and present. From flood control and transport systems to reservoirs of water, food, and energy, they are key resources for human society. To protect and sustain all of these values will require policy decisions and action at a global level, as well as ongoing advances in lake science and local management practices, and attention to the integrative approach that is the hallmark of 'limnology'.

Further reading

Historical and literary

G. Bachelard, *Water and Dreams: An Essay on the Imagination of Matter* (Dallas: The Pegasus Foundation, 1983).

C. Bertola, *Léman Maniac* [*Crazy about Lake Geneva*] (Nyon: Éditions Glénat, 2009).

J. Dennis, *The Living Great Lakes: Searching for the Heart of the Inland Seas* (New York: Thomas Dunne Books, 2003).

D. Egan, *The Death and Life of the Great Lakes* (New York: W. W. Norton & Company, 2017).

F. N. Egerton, 'History of Ecological Sciences, Part 50: Formalizing Limnology, 1870s to 1920s', *The Bulletin of the Ecological Society of America* 95(2): 131–53 (2014).

F. A. Forel, 'Notice sur l'histoire naturelle du lac Léman' [Notes on the Natural History of Lake Geneva], pp. 217–43 in: E. Rambert, H. Lebert, Ch. Dufour, F. A. Forel, and S. Chavannes (eds), *Montreux* (Neuchâtel: H. Furrer, 1877).

F. A. Forel, 'Allgemeine Biologie eines Suesswassersees' ['General Biology of a Freshwater Lake'], pp. 1–26 in: O. Zacharias (ed.), *Die Tier- und Pflanzenwelt des Suesswassers* [*The Flora and Fauna of Freshwaters*] (Leipzig: J. J. Weber, 1891).

F. A. Forel, *Le Léman: Monographie limnologique* [*Lake Geneva: Limnological Monograph*], Vols I, II, III (Lausanne: F. Rouge & Compagnie, 1892, 1895, 1904).

F. D. C. Forel (ed.), *Forel et le Léman: Aux sources de la limnologie* [*Forel and Lake Geneva: To the Origins of Limnology*] (Lausanne: Presses Polytechniques et Universitaires Romandes, 2012).

J. B. Gidmark, *Encyclopedia of American Literature of the Sea and Great Lakes* (Westport: Greenwood Press, 2001).

W. Grady (ed.), *Dark Waters Dancing to a Breeze: A Literary Companion to Rivers and Lakes* (Vancouver: Greystone Books, 2007).

B. Green, *Water, Ice & Stone: Science and Memory on the Antarctic Lakes* (New York: Harmony Books, 1995). A captivating, insightful account of lake science in the field.

J. Hart, *Storm over Mono: The Mono Lake Battle and the California Water Future* (Berkeley: University of California Press, 1996). This book has inspired students to become environmental scientists.

J. Kirk, *In the Domain of the Lake Monsters* (Toronto: Key Porter Books, 1998).

R. L. Lindeman, 'Seasonal Food-Cycle Dynamics in a Senescent Lake', *American Midland Naturalist* 1: 636–73 (1941).

S. Plath, *Crossing the Water* (London: Faber & Faber, 1975).

A. W. Reed, *Treasury of Maori Folklore* (Wellington: A. H. & A. W. Reed, 1963).

A. Steleanu, *Geschichte der Limnologie und ihrer Grundlagen* [*History of Limnology and its Foundations*] (Frankfurt: Haag & Herchen, 1989).

S. Tesson, *The Consolations of the Forest: Alone in a Cabin on the Siberian Tundra* (New York: Rizzoli International Publications Inc., 2013). A modern-day *Walden* set at Lake Baikal, Russia.

A. Thienemann, 'Seetypen' ['Lake Types'] *Naturwissenschaften* 9: 343–6 (1921).

H. D. Thoreau, *Walden* (New Haven: Yale University Press, 2006). This fully annotated, affordable version of Thoreau's 1854 classic is edited by Jeffrey S. Cramer, curator of the Thoreau Institute.

G. Topping (ed.), *Great Salt Lake: An Anthology* (Logan: Utah State University Press, 2003).

M. Twain, *Roughing It* (New York: Harper and Brothers, 1872). Includes entertaining accounts of Mark Twain's visits to Lake Tahoe and Mono Lake.

W. F. Vincent and C. Bertola, 'Lake Physics to Ecosystem Services: Forel and the Origins of Limnology', *Limnology and Oceanography e-Lectures*, 4(3), doi:10:4319/lol.2014.wvincent.cbertola.8 (2014). Available at: <http://www.cen.ulaval.ca/warwickvincent/PDFfiles/303-Forel.pdf>.

Popular guides to lake science and aquatic biology

M. J. Burgis and P. Morris, *The World of Lakes: Lakes of the World* (Ambleside: Freshwater Biological Association, 2007).

D. Gilpin and J. Schmid-Araya, *The Illustrated World Encyclopedia of Freshwater Fish & River Creatures* (London: Hermes House, 2009).

U. Lemmin, *Voyage dans les abysses du Léman* [*Voyage into the Abyssal Depths of Lake Geneva*] (Lausanne: Presses Polytechniques et Universitaires Romandes, 2016).

B. Moss, *Ponds and Small Lakes: Microorganisms and Freshwater Ecology* (Exeter: Pelagic Publishing, 2017).

L.-H. Olsen, J. Sunesen, and B. V. Pedersen, *Small Freshwater Creatures* (Oxford: Oxford University Press, 2001).

G. K. Reid et al., *Pond Life: A Guide to Common Plants and Animals of North American Ponds and Lakes* (New York: St Martin's Press, 2001).

D. W. Schindler and J. R. Vallentyne, *The Algal Bowl: Overfertilization of the World's Freshwaters and Estuaries* (Edmonton: The University of Alberta Press, 2008).

Scientific books

J. L. Awange and O. Ong'ang'a, *Lake Victoria: Ecology, Resources, Environment* (Heidelberg: Springer, 2006).

T. D. Brock, *A Eutrophic Lake: Lake Mendota, Wisconsin* (New York: Springer Verlag, 1985).

C. Brönmark and L.-A. Hansson, *The Biology of Lakes and Ponds* (Oxford: Oxford University Press, 2005).

G. A. Cole and P. E. Weihe, *Textbook of Limnology* (Long Grove: Waveland Press, 2016).

W. K. Dodds and M. R. Whiles, *Freshwater Ecology: Concepts and Environmental Applications of Limnology*, 2nd edn (San Diego: Academic Press, 2010).

S. I. Dodson, *Introduction to Limnology* (New York: McGraw-Hill, 2005).

J.-C. Druart and G. Balvay, *Le Léman et sa vie microscopique* [*Lake Geneva and its Microscopic Life*] (Versailles: Éditions Quae, 2007).

A. J. Horne and C. R. Goldman, *Limnology* (New York: McGraw-Hill, 1994).

J. Kalff, *Limnology: Inland Water Ecosystems* (Upper Saddle River: Prentice Hall, 2002).

G. E. Likens (ed.), *Encyclopedia of Inland Waters*, 3 volumes (Oxford: Elsevier, 2009).

B. Moss, *Ecology of Freshwaters: A View for the Twenty-First Century*, 4th edn (Oxford: Wiley-Blackwell, 2010).

S. T. Ross, *Ecology of North American Freshwater Fishes* (Berkeley: University of California Press, 2013).

J. P. Smol, *Pollution of Lakes and Rivers: A Paleoenvironmental Perspective*, 2nd edn (New York: John Wiley & Sons, 2008).

R. W. Sterner and J. J. Elser, *Ecological Stoichiometry: The Biology of Elements from Molecules to the Biosphere* (Princeton: Princeton University Press, 2002).

J. H. Thorp and A. P. Covich (eds), *Ecology and Classification of North American Freshwater Invertebrates*, 3rd edn (Oxford: Elsevier, 2010).

J. G. Tundisi and T. M. Tundisi, *Limnology* (Boca Raton: CRC Press, 2012).

W. F. Vincent and J. Laybourn-Parry (eds), *Polar Lakes and Rivers: Limnology of Arctic and Antarctic Aquatic Ecosystems* (Oxford: Oxford University Press, 2008).

J. D. Wehr, R. G. Sheath, and J. P. Kociolek (eds), *Freshwater Algae of North America: Ecology and Classification* (San Diego: Elsevier, 2015).

R. G. Wetzel, *Limnology: Lake and River Ecosystems*, 3rd edn (New York: Academic Press, 2001).

Scientific articles and reviews

S. Bonilla and F. R. Pick, 'Freshwater Bloom-Forming Cyanobacteria and Anthropogenic Change', *Limnology and Oceanography e-Lectures* 7(2) (2017), <https://doi.org/10.1002/loe2.10006>.

J. Catalan et al., 'High Mountain Lakes: Extreme Habitats and Witnesses of Environmental Changes', *Limnetica* 25: 551–84 (2006).

B. C. Christner, J. C. Priscu, et al., 'A Microbial Ecosystem beneath the West Antarctic Ice Sheet', *Nature* 512: 310–13 (2014).

J. J. Cole et al. 'Plumbing the Global Carbon Cycle: Integrating Inland Waters into the Terrestrial Carbon Budget', *Ecosystems* 10: 172–85 (2007).

B. R. Deemer et al., 'Greenhouse Gas Emissions from Reservoir Water Surfaces: A New Global Synthesis', *BioScience* 66: 949–64 (2016).

J. A. Downing, 'Emerging Global Role of Small Lakes and Ponds', *Limnetica* 29: 9–24 (2010).

D. Dudgeon et al. 'Freshwater Biodiversity: Importance, Threats, Status and Conservation Challenges', *Biological Reviews* 81(2): 163–82 (2006).

L. Grattan and V. Trainer (eds), 'Harmful Algal Blooms and Public Health', *Harmful Algae* 57B: 1–56 (2016).

F. Hölker et al., 'Tube-Dwelling Invertebrates: Tiny Ecosystem Engineers have Large Effects in Lake Ecosystems', *Ecological Monographs* 85(3): 333–51 (2015).

S. MacIntyre and R. Jellison, 'Nutrient Fluxes from Upwelling and Enhanced Turbulence at the Top of the Pycnocline in Mono Lake, California', *Hydrobiologia* 466: 13–29 (2001).

M. V. Moore et al., 'Climate Change and the World's "Sacred Sea"—Lake Baikal, Siberia', *BioScience* 59: 405–17 (2009).

R. J. Newton et al., 'A Guide to the Natural History of Freshwater Lake Bacteria', *Microbiology and Molecular Biology Reviews* 75: 14–49 (2011).

C. M. O'Reilly et al., 'Rapid and Highly Variable Warming of Lake Surface Waters around the Globe', *Geophysical Research Letters* 42(24) (2015).

H. W. Paerl et al., 'It Takes Two to Tango: When and Where Dual Nutrient (N & P) Reductions are Needed to Protect Lakes and Downstream Ecosystems', *Environmental Science & Technology* 50: 10805–13 (2016).

L. G. M. Pettersson, R. H. Henchman, and A. Nilsson, 'Water: The Most Anomalous Liquid', *Chemical Reviews* 116: 7459–62 (2016).

S. Pointing et al., 'Quantifying Human Impact on Earth's Microbiome', *Nature Microbiology* 1: 16145 (2016).

D. Righton et al., 'Empirical Observations of the Spawning Migration of European Eels: The Long and Dangerous Road to the Sargasso Sea', *Science Advances* 2: e1501694 (2016).

J. P. Smol, 'Paleolimnology: An Introduction to Approaches Used to Track Long-Term Environmental Changes Using Lake Sediments', *Limnology and Oceanography e-Lectures* 1(3) (2009), <https://doi.org/10.4319/lol.2009.jsmol.3>.

J. A. Stenson, 'Differential Predation by Fish on Two Species of *Chaoborus* (Diptera, Chaoboridae)', *Oikos* 31: 98–101 (1978).

J. D. Stockwell et al., 'Habitat Coupling in a Large Lake System: Delivery of an Energy Subsidy by an Offshore Planktivore to the Nearshore Zone of Lake Superior', *Freshwater Biology* 59: 1197–212 (2014).

C. A. Suttle, 'Environmental Microbiology: Viral Diversity on the Global Stage', *Nature Microbiology* 1: 16205 (2016).

C. S. Turney and H. Brown, 'Catastrophic Early Holocene Sea Level Rise, Human Migration and the Neolithic Transition in Europe', *Quaternary Science Reviews* 26: 2036–41 (2007).

C. Verpoorter et al., 'A Global Inventory of Lakes Based on High-Resolution Satellite Imagery', *Geophysical Research Letters* 41: 6396–402 (2014).

C. E. Williamson et al., 'Ecological Consequences of Long-Term Browning in Lakes', *Scientific Reports* 5: 18666 (2015).

K. O. Winemiller et al., 'Balancing Hydropower and Biodiversity in the Amazon, Congo, and Mekong', *Science* 351: 128–9 (2016).

Websites

ASLO: <http://aslo.org>.
Billow demonstration (Kelvin–Helmholtz instabilities): <https://www.youtube.com/watch?v=UbAfvcaYr00>
Chironomid tubes (video): <https://www.youtube.com/watch?v=RQwau_uSyy4>
Colour of lakes: <http://www.citclops.eu/transparency/measuring-water-transparency>
Cyanobacteria-identification: <https://pubs.usgs.gov/of/2015/1164/ofr20151164.pdf>
Dams (data base): <http://www.icold-cigb.org/GB/world_register/world_register_of_dams.asp>
Daphnia feeding (video): <https://www.youtube.com/watch?v=pLL_YzZ_4O0>
Daphnia swimming (video): <https://www.youtube.com/watch?v=MyDQ_f1mzH8>
English Lake District and other U.K. lakes: <https://eip.ceh.ac.uk/apps/lakes/>
Kelvin waves: <https://www.youtube.com/watch?v=SZlix47Jq4A>
Lake Biwa: <http://www.biwahaku.jp/english/member-e/researchactivities.html>
Lake Tahoe: <http://terc.ucdavis.edu/>
Large lakes (IAGLR site): <http://www.iaglr.org/lakes/>

Léman/Lake Geneva—International Commission (CIPEL) <http://www.cipel.org/>

Methane from Arctic thaw lakes (video): <https://www.dailymotion.com/video/x2mwrcv>

Microbial mats in Lake Untersee, Antarctica (video): <https://www.youtube.com/watch?v=qs2hUZP-6Bo>

Mono Lake: <http://www.monolake.org/>

NALMS: <https://www.nalms.org/>

Phantom midge (video): <https://www.youtube.com/watch?v=LQCj6T5sdQM>

Plankton and benthos (images from the Freshwater Biological Association): <http://www.environmentdata.org/browse-collection>

Rotifers (images): <http://www.microscopy-uk.org.uk/mag/wimsmall/extra/rotif.html>

SIL: <http://limnology.org/>

Tubifex worms—sludge worms (video): <https://www.youtube.com/watch?v=hxYBiBi3EbE>

World Lake Database (reference catalogue): <http://wldb.ilec.or.jp/>

"牛津通识读本"已出书目

古典哲学的趣味
人生的意义
文学理论入门
大众经济学
历史之源
设计，无处不在
生活中的心理学
政治的历史与边界
哲学的思与惑
资本主义
美国总统制
海德格尔
我们时代的伦理学
卡夫卡是谁
考古学的过去与未来
天文学简史
社会学的意识
康德
尼采
亚里士多德的世界
西方艺术新论
全球化面面观
简明逻辑学
法哲学：价值与事实
政治哲学与幸福根基
选择理论
后殖民主义与世界格局

福柯
缤纷的语言学
达达和超现实主义
佛学概论
维特根斯坦与哲学
科学哲学
印度哲学祛魅
克尔凯郭尔
科学革命
广告
数学
叔本华
笛卡尔
基督教神学
犹太人与犹太教
现代日本
罗兰·巴特
马基雅维里
全球经济史
进化
性存在
量子理论
牛顿新传
国际移民
哈贝马斯
医学伦理
黑格尔

地球
记忆
法律
中国文学
托克维尔
休谟
分子
法国大革命
丝绸之路
民族主义
科幻作品
罗素
美国政党与选举
美国最高法院
纪录片
大萧条与罗斯福新政
领导力
无神论
罗马共和国
美国国会
民主
英格兰文学
现代主义
网络
自闭症
德里达
浪漫主义

批判理论	德国文学	儿童心理学
电影	戏剧	时装
俄罗斯文学	腐败	现代拉丁美洲文学
古典文学	医事法	卢梭
大数据	癌症	隐私
洛克	植物	电影音乐
幸福	法语文学	抑郁症
免疫系统	微观经济学	传染病
银行学	湖泊	